ß

D1760103

Geology of the coastal region between Lowestoft and Saxmundham

This memoir describes the sediments and rocks of the coastal region of East Anglia between Lowestoft and Aldeburgh, outlines the history of their formation, and discusses their relationship to the present day landscape. The local geology has greatly influenced many aspects of human activity in the area, including agriculture, industry and tourism.

The district, with its broad plateaux, valleys, and wide flat marshlands, carries the imprint of geological and climatic events from the last 500 000 years or so. First an ice sheet originating in northern Britain advanced across East Anglia depositing chalky stony clay, whilst meltwaters issuing from the ice produced spreads of sand and gravel. At about the same time another ice sheet originating from Scandinavia advanced across northern East Anglia and deposited another set of clays, sands and gravels. The ice sheets then retreated. Subsequent periods of alternating temperate and cold climate were associated mainly with erosion, but included times when river terraces were deposited. Then, about 13 000 years ago, melting of a later ice sheet caused the sea level to rise. This ice sheet did not extend as far south as the present district. The coastal valleys were flooded and silted up, and peat formed around the marshes. The peat was subsequently dug, in historical times, and the workings became flooded to produce the famous 'Broads'.

Deep beneath the district, slates of Silurian age (about 400 million years old), have been proved in a borehole at Lowestoft. These form part of a basement massif that underlies the whole of south-eastern Britain. A major stratigraphical break separates the slates from the overlying Cretaceous rocks, which include the well-known Chalk, the region's main aquifer. Overlying the Chalk is a sequence of Palaeogene clays recorded in several boreholes. The succeeding deposits of Crag sands and gravels crop out extensively and are also well exposed in cliffs, and in several quarries, within the district.

Coastal erosion is a major problem in this district. A good understanding of the local geology is an essential prerequisite to effective remedial action.

Cover photograph   Coastline looking north from Dunwich Heath [TM 478 677] (Photographer: R J O Hamblin) (GS 558).

Plate 1   North-south cliffs at Dunwich Heath showing large south–east-dipping cross-sets of well-rounded flint gravels (Westleton Beds) interbedded with sands of the Norwich Crag Formation. (GS 557)

BRITISH GEOLOGICAL SURVEY

B S P MOORLOCK
R J O HAMBLIN
S J BOOTH
A N MORIGI

Geology of the country around Lowestoft and Saxmundham

Memoir for 1:50 000 Geological Sheets 176 and 191 (England and Wales)

CONTRIBUTORS

Stratigraphy
P S Balson
P M Hopson
D McC Bridge
D H Jeffery
S J Mathers
J A Zalasiewicz

Hydrogeology
M A Lewis

Petrology
C R Hallsworth

Palaeontology
M A Woods
I P Wilkinson
R Harland

Geophysics and Deep geology
J D Cornwell
T C Pharaoh

Economic geology
P M Harris

Database
R T Mogdridge

London: The Stationery Office   2000

ISBN   0 11 884543 8

Bibliographical reference

MOORLOCK, B S P, HAMBLIN, R J O, BOOTH, S J, and MORIGI, A N. 2000.   Geology of the country around Lowestoft and Saxmundham.   *Memoir of the British Geological Survey*, Sheets 176 and 191 (England and Wales).

Authors

B S P Moorlock, BSc, PhD
R J O Hamblin, BSc, PhD, CGeol
S J Booth, BSc
A N Morigi, BSc
British Geological Survey, Keyworth

Contributors

P S Balson, BSc, PhD
D McC Bridge, BSc
C R Hallsworth, BSc
P M Hopson, BSc, CGeol
S J Mathers, BSc
R T Mogdridge
T C Pharaoh, BSc, PhD, CGeol
I P Wilkinson, MSc, PhD, CGeol
M A Woods, BSc
British Geological Survey, Keyworth

M A Lewis, BA, MSc
British Geological Survey, Wallingford

J M Allsop, BSc, CGeol, MIScT
J D Cornwell, MSc, PhD
R Harland, DSc
P M Harris, MA, CGeol, CEng
A Horton, BSc
D H Jeffery, BSc, MSc, CGeol
J A Zalasiewicz, BSc, PhD
formerly British Geological Survey

Printed in the UK for The Stationery Office
TJ 001711   C6   3/2000

Other publications of the Survey dealing with this and adjoining districts

BOOKS
British Regional Geology
East Anglia and adjoining areas (4th edition), 1982
Memoirs
Geology of the country around Norwich (Sheet 161), 1982
Geology of the country around Great Yarmouth (Sheet 162), 1994
Geology of the country around Diss (Sheet 175), 1993
Mineral Assessment Report (sand and gravel resources)
Harleston and Bungay, 1985
Offshore Reports
The geology of the Southern North Sea, 1992
Well catalogues
Records of wells in the area of the New Series one-inch Geological Sheet 161 (Norwich)
Records of wells in the area of the New Series one-inch Geological Sheet 162 (Great Yarmouth)

MAPS
1:625 000
Solid geology (south sheet), 1979
Quaternary geology (south sheet), 1977
Aeromagnetic anomaly (south sheet), 1965
Bouguer gravity anomaly (south sheet), 1986
1:250 000
52N 00 East Anglia
  Solid geology, 1986
  Quaternary geology, 1991
  Aeromagnetic anomaly, 1982
  Bouguer gravity anomaly, 1981
1:125 000
Hydrogeology, northern East Anglia, 1976
Hydrogeology, southern East Anglia , 1981
1:50 000 (Solid and Drift)
Sheet 161 (Norwich), 1975
Sheet 175 (Diss), 1989
Sheet 162 (Great Yarmouth), 1990
Sheet 189 (Bury St Edmunds), 1982
Sheet 207 (Ipswich), 1990
Sheet 208/225 (Woodbridge and Felixstowe), 1977

CONTENTS

One   Introduction   1
Location and topography   1
Outline of geological history   1
Previous research   4

Two   Concealed strata: Silurian to Palaeogene   6
Lower Palaeozoic basement   6
Mesozoic rocks   9
 Cretaceous   9
  Lower Cretaceous   10
  Upper Cretaceous   10
   Chalk Group   10
Palaeogene rocks   13
 Nomenclature   14
 Palaeocene–Eocene boundary   14
  Ormesby Clay Formation   14
 Lambeth Group   16
 Thames Group   18
  Harwich Formation   18
  London Clay Formation   21
 Depositional sequence of Palaeogene deposits   21
 Magnetic susceptibility   21
 Clay mineralogy   23

Three   Late Neogene and Early Quaternary: The Crags and related deposits   24
Coralline Crag Formation   24
Crag Group   26
Cromer Forest-bed Formation   42
Kesgrave Group   44
Bytham Sands and Gravels   47
'Beccles Beds'   51

Four   Quaternary (Anglian to Devensian): glacial and interglacial deposits   53
Corton Formation   53
Lowestoft Till Formation   61
 Oulton Beds   64
 Corton Woods Sands and Gravels   65
 Haddiscoe Sands and Gravels   65
 Aldeby Sands and Gravels   66
 Unnamed sands and gravels   67
Interglacial deposits   67
 Hoxnian interglacial deposits   67
 Ipswichian interglacial deposits   68
River terrace deposits   68
 Waveney valley   68
 River Blyth   71
 River Deben   71
 River Alde   71
Yare Valley Formation   71
Head   72
Older blown sand   73

Five   Quaternary (Holocene): postglacial and present-day deposits   74
Marine deposits   76
Holocene valley deposits   79
 Waveney valley: Breydon Formation   79
 Other valleys   82
Blown sand   83
Ground modified by human activity   83

Six   Structure   85
Carboniferous to early Mesozoic tectonism   85
Cretaceous and later tectonic events   85

Seven   Economic geology   87
Mineral deposits   87
 Sand and gravel   87
 Brick clay   87
 Peat   88
Hydrogeology and water supply   88
 Chalk aquifer and overlying Palaeogene deposits   89
 Crag Group aquifer and overlying deposits   93
Engineering geology   94
Coastal erosion   95

References   97

Appendices   107
1   Geological Survey photographs   107
2   Abstracts of selected borehole logs   108
3   History of survey   110

Fossil index   111

General index   112

FIGURES

1   Topography of the district   2
2   Solid geology of the district   3
3   Distribution of boreholes proving basement in East Anglia   6
4   Gravity and magnetic anomaly maps, main lineaments and model of crustal structure   8
5   Transient electromagnetic sounding for a site near Snape   9
6   Schematic section showing the principal unconformities affecting the Palaeozoic and Mesozoic strata abutting the London–Brabant Massif   10
7   Cretaceous deposits proved in boreholes within the district   11
8   Thickness variations of the Palaeogene formations   13
9   Generalised palaeogeography of north-west Europe during the early Palaeogene   14

10 Location of important boreholes proving Palaeogene deposits in the district and adjacent areas 18

11 Correlation of Palaeogene strata between East Anglia and Kent 19

12 Magnetic susceptibility of cored material from boreholes 22

13 Regional distribution of the Crag Group and Coralline Crag 24

14 Relationships between the lithostratigraphical components of the Crag Group 28

15 Contours on the base of the Crag Group 29

16 Sketch sections of Norwich Crag gravels (Westleton Beds) and measurement of large-scale cross-stratification 34

17 Geological cross-section from Aldeburgh to North Warren 43

18 Approximate limits of the Kesgrave Group, Bytham Sands and Gravels, Cromer Forest-bed Formation and Corton Formation in the district 46

19 Bytham Sands and Gravels in the district 49

20 The generalised succession at Corton 54

21 Variation of the Corton Formation proved in boreholes in the western part of the district 60

22 Schematic cross-section of the Lowestoft Till and Corton formations in the north-east of the district 62

23 Anglian meltwater channels 63

24 River terraces of the Waveney 69

25 Distribution of alluvium, tidal flat deposits and Breydon Formation in the district 75

26 Contours on the base of the Breydon Formation within the Waveney valley 76

27 Schematic section showing the relationships of the Holocene deposits from the hinterland to the coast and the nearshore 77

28 Schematic section through the upper Waveney valley showing the limits of the Breydon Formation 80

29 Waveney valley: distribution of soils liable to acidify if deep-drained 81

30 Map showing contours in metres on the minimum potentiometric surfaces of the Chalk and Crag Group 91

TABLES

1 Main developments in the classification of the early Palaeogene of south-eastern England 15

2 Lithostratigraphical classification and correlation of early Palaeogene strata of the London Basin, East Anglia and the Southern North Sea Basin 16

3 Application of the magnetic and biostratigraphical record to the Palaeogene deposits of south-east England 17

4 An outline of the nomenclature of the Crag Group and associated strata 27

5 Correlation of lithostratigraphy between the present district and the Southern North Sea 38

6 Clast lithological analysis of samples from the Bytham Sands and Gravels, with comparative analyses from the Corton and Cromer Forest-bed formations 45

7 Lithostratigraphical nomenclature adopted for the Anglian glacial deposits of the district, compared with a selection of earlier schemes 55

8 Composition of the two gravels at Flixton Quarry 70

9 Classification of marshland sediments in the Waveney valley 79

10 Quantity of water licensed to be abstracted annually from the Lowestoft and Saxmundham district 89

11 Chemical analyses of groundwater from the main aquifers within or near the district 90

PLATES

Cover photograph Coastline looking north from Dunwich Heath

1 *Frontispiece* North-south cliffs at Dunwich Heath showing large SE-dipping cross-sets of well-rounded flint gravels (Westleton Beds) interbedded with sands of the Norwich Crag Formation

2 Lambeth Group lithologies in the Halesworth Borehole 20

3 Clay within the Norwich Crag Formation exposed as a bench on the foreshore at Covehithe; overlain by Norwich Crag sands and Corton Formation 30

4 Detail of clay within the Norwich Crag Formation at Covehithe, showing bioturbation 31

5 Well-rounded flint gravels (Westleton Beds) within the Norwich Crag Formation at Blyth River Gravel Pit near Halesworth 32

6 Seam of clay within well rounded flint gravels (Westleton Beds) exposed at Thorington Gravel Pit 35

7 Sands and interbedded thin clays within the Norwich Crag Formation, Covehithe cliffs 39

8 Cross-bedded sands of the Corton Formation overlain by Lowestoft till 56

9 Drying shed at Cove Bottom Brickworks, near Wrentham 88

10 Cliff erosion at Pakefield 96

PREFACE

This memoir and its two accompanying 1:50 000 scale geological maps identify the geological events that have produced the present landscape and its coastline, and demonstrate the importance of the geology of the district to a wide range of environmental and socio-economic issues.

There is great concern at present over the effect global warming and associated sea-level rise that might be caused by human activities. For this reason it is very important that we are able to separate natural fluctuation of sea level from those related to human activity. The geological record is critical to doing this and for example study of the geology of the district reveals that natural processes have caused the sea level to fluctuate greatly during the last two million years; at times it has been much higher than at present, at other times much lower.

During this last two million years central East Anglia has been subjected to a wide range of climatic and environmental changes, many of which are reflected in the underlying geological deposits. Initially the sea encroached over the area from the east, depositing sands and local gravels, the latter interpreted as fossil beach deposits. The sea then retreated and the southern part of the district was crossed by an extensive braided river system, considered to be the forerunner of the River Thames, and this deposited another set of sands and gravels. In the north there is evidence to indicate the presence of a second early river, this one flowing eastwards from the English Midlands. Slightly later, two large ice-sheets extended into the area, one originating in Scandinavia and the other in northern Britain. They left behind widespread sheets of sand, gravel and clay, including the distinctive chalky boulder clay (till) which remains to the present day over much of the higher ground of East Anglia and the East Midlands. During the glacial maximum the sea level was probably 100 m or so below its present level, much of the earth's oceanic water having been locked up as ice. Within the last 8000 years a general global rise in sea level has drowned the river valleys so that silts, clays and peats have accumulated in them. Severe coastal erosion over the last few centuries has resulted in the loss of the former town of Dunwich and several villages, but elsewhere within the district sands and gravels are accreting along the shoreline and land is being gained from the sea.

This memoir describes the sediments and rocks underlying central East Anglia, from the port of Lowestoft southwards to Aldeburgh, and inland as far west as the small towns of Bungay and Framlingham. The oldest rocks, metamorphosed mudstones of Silurian age, form a basement massif known from deep boreholes. A large unconformity separates these rocks from the overlying Cretaceous deposits, which are present at depth beneath the entire district. In the east of the district the Upper Chalk, of Upper Cretaceous age, is overlain unconformably by Palaeogene clays. The Halesworth Borehole, drilled during the survey, has provided important new stratigraphical information on these deposits, which were laid down along the western margin of the Palaeogene North Sea Basin. The Palaeogene clays are overlain by the Crag Group, a sequence of predominantly sandy beds including local gravels and clays. The Crag is regarded as the youngest of the Solid strata; overlying glacial and later deposits are classified as Drift.

East Anglia is an area prone to coastal flooding and any rise in sea level whether related to a global rise or a relative rise resulting from a lowering of the land area, will exacerbate this problem. This study does not of itself provide any solution to this problem but it does provide the scientific base from which the planners and the decision makers can start to develop potential solutions.

David A Falvey, PhD
Director

British Geological Survey
Kingsley Dunham Centre
Keyworth
Nottingham
NG12 5GG

ACKNOWLEDGEMENTS

NOTES

The memoir was written by B S P Moorlock, R J O Hamblin, S J Booth and A N Morigi, and incorporates a compilation of published and unpublished work by P M Hopson, S J Booth, A N Morigi, R J O Hamblin, B S P Moorlock, A Horton, J A Zalazeiwicz, S J Mathers, P S Balson, T E Lawson and C Wilcox. Geophysical data and intrepretation in Chapter 2 were provided by J D Cornwell, whilst the section on the Palaeozoic basement is largely the work of T C Pharaoh.

P S Balson wrote the sections on offshore geology and the Coralline Crag. M A Woods provided information on the Chalk. In Chapter 7 the hydrogeological account of the district was written by M A Lewis, and the mineral deposits were described by P M Harris.

Micropalaeontological identifications are by R Harland and I P Wilkinson. Heavy mineral studies were made by C R Hallsworth. Detailed logging of the Halesworth Borehole was undertaken by A N Morigi. S J Booth was responsible for organising and developing the computer database of subsurface information. Much of the borehole information for this database was input by R T Mogdridge. The photographs were taken by the field survey team. The memoir was edited by J I Chisholm.

This work has benefited from the generous assistance afforded by Professor B M Funnell (University of East Anglia) and Professor J Rose (Royal Holloway, University of London) in respect of the Quaternary stratigraphy. There has also been fruitful collaboration with the Soil Survey of England and Wales (now the Soil Survey and Land Research Centre, Cranfield University), who have provided unpublished data on the distribution of peat in the Alde and Blyth valleys.

We gratefully acknowledge the willing cooperation of many public and private bodies in supplying borehole and survey information, and arranging for our access to construction sites, quarries and nature reserves during the survey; these bodies include Anglian Water Service Ltd., Essex and Suffolk Water, National Rivers Authority (Anglia Region), Suffolk County Council, English Nature, The Soil Survey and Land Research Centre, The Ministry of Agriculture, Fisheries and Food, University of East Anglia (Professor B Funnell), ARC (Eastern), Atlas Aggregates, Minns Aggregates, Cove Bottom Brickworks, Port of Lowestoft Authority, the Forestry Commission, Heritage Coast District Council, Waveney District Council, Southwold Borough Council, and Reynolds Hardiman and Partners. Finally, we acknowledge the ready cooperation of the many landowners, tenants and wardens of the district, who freely gave us access to their properties during the course of the surveys.

Throughout the memoir the word 'district' refers to the area (including the offshore area) covered by the 1:50 000 Lowestoft (176) and Saxmundham (191) geological map sheets.

References to 'the published map' are specifically to the 1:50 000 Lowestoft (176) and Saxmundham (191) sheets, including the geological sections and other marginal information shown thereon.

National Grid references are given in square brackets; they lie within 100 km square TM unless otherwise stated.

Numbers prefixed by 'E' refer to the National Sliced Rock Collection of the British Geological Survey, housed at Keyworth, Nottingham.

ONE

Introduction

LOCATION AND TOPOGRAPHY

This memoir describes the geology of the district covered by two adjacent sheets of the 1:50 000 Geological Map of England and Wales, namely Lowestoft (Sheet 176) and Saxmundham (Sheet 191). The district extends eastwards to include the nearshore zone of the North Sea.

The district lies predominantly within the county of Suffolk but includes a small part of southern Norfolk to the north of the Waveney valley. Much of the area consists of a dissected plateau, which rises some 50 m above OD in the west, but decreases gently to the east where the uplands become separated by valleys and tracts of marshland (Figure 1). In the north this low ground forms part of the extensive system of wetlands, broads and drained marshes known as Broadland. The northern and central parts of the district are drained by the rivers Blyth, Yox and Waveney, all of which flow generally to the east, although the last-named turns abruptly northwards before discharging into the sea north of Lowestoft. The southern part of the district lies within the catchments of the rivers Alde and Deben and their tributaries.

The town of Lowestoft was once a major fishing port, but with the fishing industry in decline it now relies heavily on tourism. Southwold, some 17 km to the south, developed as a holiday resort in Victorian and Edwardian times and has remained popular with tourists ever since. Saxmundham, Framlingham, Beccles and Bungay function as market towns. Aldeburgh is renowned for its annual arts festival.

The extensive wetlands along the coastal fringe attract a large variety of birds, making the district popular with ornithologists. The very local stone curlew and marsh harrier both breed within the area. Minsmere, perhaps the best known of the reserves maintained by the Royal Society for the Protection of Birds, lies near the coast some 10 km to the south of Southwold. The coastline here is dominated by the recent extensive development of the Sizewell Power Station.

Much of the plateau area is given over to the production of cereals, although sugar beet and oil seed rape are also grown extensively. Peas, field beans, brussel sprouts, parsnips and potatoes are also produced locally, mainly on the more sheltered valley sides. Cattle are reared on many of the farms, particularly those with access to drained marshland where they can be pastured during the summer months. Sheep and pigs are also farmed, the latter in large units scattered across the district.

Historically the marshlands have supplied peat, and the uplands have provided sand, gravel, marling clay and brick-clay. Several sand and gravel pits were operating within the district at the time of survey. At Cove Bottom, clay from the Crag is dug on a limited scale for use in the manufacture of hand-made bricks. Similar clay was also dug and made into bricks at Aldeburgh, but the clay is now brought to the brickworks from elsewhere. Abandoned excavations, on all scales, have been utilised for landfill, ranging from the local dumping of domestic and agricultural waste in small pits to large-scale disposal under local authority control.

The Chalk provides about half of the district's groundwater requirements; the remainder is obtained in about equal proportions from the Crag and the drift deposits. In recent years an increasing amount of water has been used for irrigation.

Erosion is, and has been for many centuries, a major problem along much of the coastline; most of the losses occur in the winter months during periods of high rainfall, storms and high tides. In the Covehithe area the cliffs have retreated at an average rate of about 5 metres per year during the last few years, although annual losses can fluctuate greatly. Farther south, the village of Dunwich, in the early fourteenth century a large wealthy seaport with many fine churches, has almost been consumed by the sea; only a small hamlet remains. Elsewhere, for example at Kessingland, just south of Lowestoft, the land is actively accreting by the addition of gravel along the shoreline.

OUTLINE OF GEOLOGICAL HISTORY

The district lies on the northern margin of the London–Brabant Massif (Wills, 1978) a concealed platform of Palaeozoic rocks which acted as a positive structural feature through Upper Palaeozoic and Mesozoic times.

The Solid geology of the district is shown in Figure 2. This indicates the distribution of Mesozoic, Palaeogene and Quaternary bedrock formations at outcrop or subcrop beneath the cover of later Quaternary drift deposits.

Palaeozoic and Mesozoic

The London–Brabant Massif consists of folded and metamorphosed sedimentary and volcanic rocks with late Caledonian intrusions. The rocks proved in the upper parts of the massif in and adjacent to the district are metasedimentary rocks of Silurian age, including cleaved mudstones, siltstones and sandstones. These are believed to belong to a turbiditic shelf succession which was folded and metamorphosed during the Acadian (early Devonian) orogeny.

The massif formed a structural high and was flanked by sedimentary basins through the Upper Palaeozoic and Mesozoic; sedimentary formations that accumulated in the adjoining Southern North Sea Basin thinned out

Figure 1
Topography of
the district.

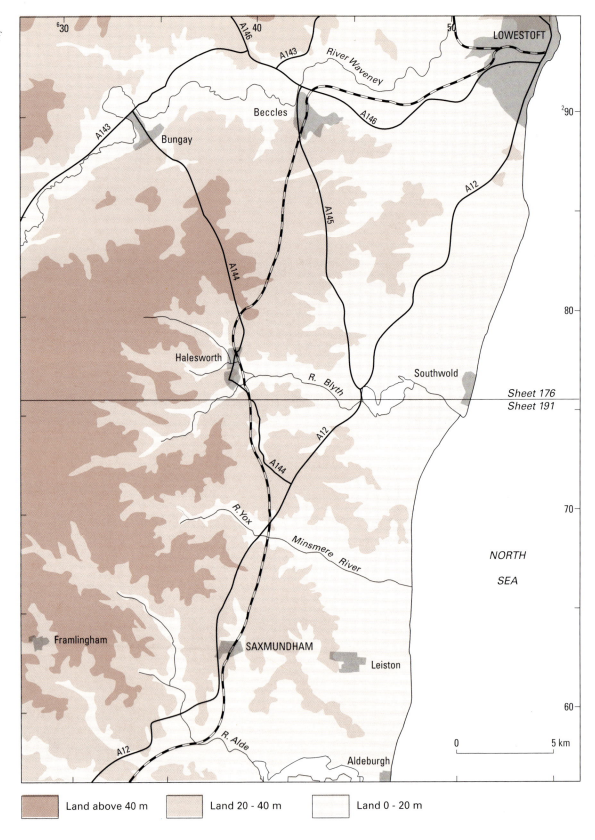

Land above 40 m   Land 20 - 40 m   Land 0 - 20 m

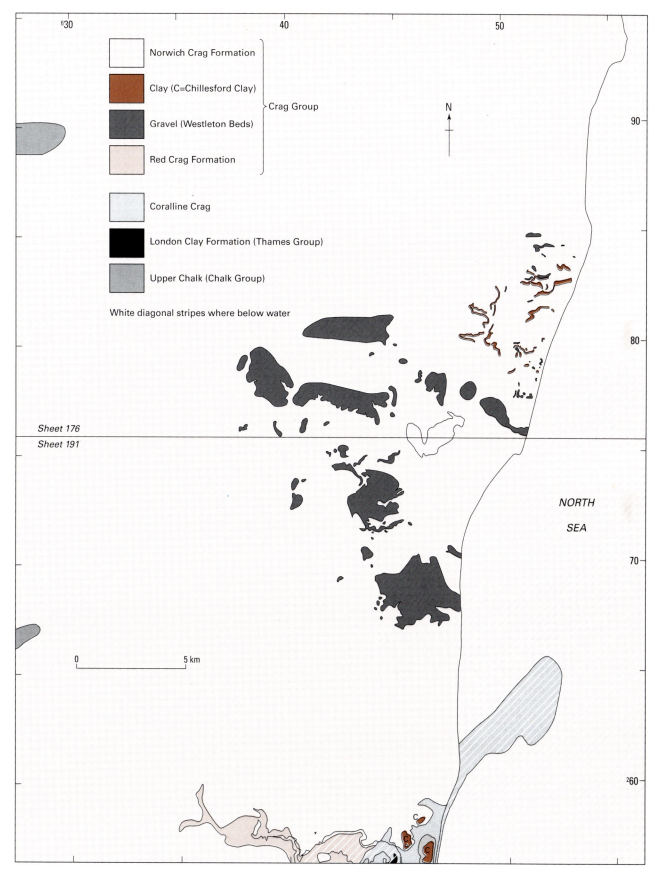

Figure 2   Solid geology of the district.

against its northern margin. Within the district there is no evidence to indicate the presence of concealed Carboniferous deposits but they are preserved some 15 km to the north, in the Great Yarmouth district.

The interaction of the relatively stable platform with the adjacent subsiding Southern North Sea Basin resulted in attenuation, non-deposition and erosion of Permian, Triassic and Jurassic strata within the region. If any of these deposits did accumulate in the district they were removed by Late-Cimmerian (late Jurassic to early Cretaceous) erosion. It was not until Albian (early Cretaceous) times that regional subsidence resulted in the complete overstep of the Palaeozoic platform (Owen, 1971). The deposition of the early Cretaceous Gault marked the start of a period of sea level rise that continued throughout most of the late Cretaceous, the surface of the massif being progressively buried, most notably by the limestones of the Chalk. The lack of ter-rigenous material within the Chalk is probably due to the low relief of the landmasses bordering the Chalk sea, and to their distant location.

Palaeogene

Regional uplift and erosion during the late Cretaceous was followed by subsidence of the Southern North Sea Basin, which continued throughout the Palaeogene and Quaternary, perhaps interrupted by uplift in mid to late Miocene times. Differential subsidence of the complex graben structures that formed during the Mesozoic as part of the North Atlantic rift system, and which had con-trolled sedimentation in the North Sea Basin, ceased to operate by early Palaeogene times. Subsequent sub-sidence formed a saucer-shaped depression along the axis of the Mesozoic graben system.

The Ormesby Clay Formation marks the transgression of the Palaeocene sea over the eroded Chalk surface. It was followed by the Lambeth Group (formerly known as the Woolwich and Reading Beds), which formed under shallow marine and estuarine conditions. A further marine transgression resulted in the deposition of the Harwich and London Clay formations of the recently defined Thames Group (Ellison et al., 1994). Layers of volcanic ash within the Thames Group are the product of volcanism associated with the contemporaneous rifting of the north Atlantic.

Late Palaeogene to the present

Deposition of the Palaeogene strata was followed by a period of uplift, and no further sediments accumulated in the district until a late Pliocene transgression deposited the Coralline Crag in the south-east. Further marine transgressions in the early Pleistocene resulted in the deposition of the Crag Group in shallow shelf and intertidal environments. These deposits are mainly sands, silts and clays, but locally include gravels. Marine condi-tions then gave way to a fluvial regime which deposited sand and gravel over a range of time perhaps extending from the Pastonian to the Beestonian. There is evidence to indicate that some of these sands and gravels were deposited by a forerunner of the River Thames which flowed along a more northerly course than the present

river; others appear to be the products of a separate eastward-flowing river system to the north of the Proto-Thames. A hiatus in sedimentation, accompanied by a period of widespread soil formation in southern East Anglia, was followed by the onset of the Anglian Stage, a period of cold climatic conditions represented over much of eastern England by glacial deposits.

The products of two Anglian ice sheets are represented within the district, one originating in Scandinavia, the other in northern Britain. Conflicting views are held as to whether the two ice sheets coexisted or were separated in time by a period of erosion. The glacial deposits include tills, outwash sands and gravels and glaciolacustrine silts and clays.

The climate warmed at the end of the Anglian and interglacial deposits of Hoxnian age are preserved in depressions on the till surface. Younger (Ipswichian age) interglacial deposits are also preserved locally lower down on the valley sides.

By Devensian times the landscape was probably very similar to that of the present day, although rivers would have been graded to a lower sea level, with the cliffline located several kilometres east of its current position. During the Devensian, much of Britain was subject once again to glaciation, but the ice sheet failed to reach this part of East Anglia. The sands and gravels of the first terrace of the River Waveney probably formed during this period.

The sea level rose during the succeeding Flandrian Stage, and broad tracts of estuarine sediments were deposited in the east of the district. The coastline also suffered major erosion and retreat. Layers of peat within the alluvial sequence mark brief regressive periods during the Flandrian.

PREVIOUS RESEARCH

In the late nineteenth century the findings of the original one-inch to one-mile survey of the region were written up in a series of memoirs by Dalton and Whitaker (1886), Reid (1890), Whitaker (1887), Whitaker and Dalton (1887) and Blake (1890). These works record a vast amount of information, and include details of a large number of exposures, both natural and man-made, that are no longer available for study. Complementing these studies was a series of papers by Harmer (1898, 1900a, 1900b, 1902, 1910a, 1910b, 1910c, 1924) on various aspects of the Palaeogene and Quaternary geology of Britain, with particular reference to eastern England.

Relatively little was then published on the geology of East Anglia until after the Second World War when Baden-Powell (1948) divided the chalky till into a lower till and an upper till, which he believed to be the products of two glaciations separated by the Hoxnian interglacial episode. This, together with the work of West (1956) on the interglacial sediments at Hoxne and the latter's subsequent review of British Pleistocene stratig-raphy (West, 1963) sparked a revival of interest in East Anglian Quaternary geology.

Baden-Powell's views were supported by West and Donner (1956) but later challenged by Bristow and Cox (1973) who argued that all the chalky till to the south of Norwich formed as a single pre-Hoxnian sheet. This was supported by Perrin et al. (1973) who demonstrated that the chalky tills of East Anglia and the East Midlands are very uniform mineralogically. Sumbler (1983) has argued further that the type Wolstonian sequence in Warwickshire and the type Anglian sequence of East Anglia were deposited in the same pre-Hoxnian stage.

The work of Funnell (1961), West (1961a, b, 1980), Funnell and West (1977), and Beck et al. (1972) revived interest in the Crag deposits. Much subsequent debate has centred on the chronostratigraphy of the Crag, with several authors (Zalasiewicz and Gibbard, 1988; Zalasiewicz et al., 1991; Gibbard et al., 1991) disputing the order of the stages erected by the earlier workers, and arguing that some of the stages have been duplicated. Bristow (1983), working in the Bury St Edmunds district, claimed that it was not possible to map boundaries within the Crag, although this was later achieved by Zalasiewicz and Mathers (1985) and Zalasiewicz et al. (1988) in the southern part of the present district.

Recent years have seen also the gradual unravelling of the mid-Pleistocene history of East Anglia. In 1976 Rose et al., identified the Kesgrave Sands and Gravels, a suite of quartz and quartzite bearing gravels, across much of southern East Anglia. These are thought to have been deposited prior to the Anglian glaciation by the ancestral River Thames, when it flowed along a more northerly course than at present. The stratigraphy of the gravels has been refined by Hey (1980), Whiteman (1992) and Whiteman and Rose (1992) who have raised the status of the gravels to a group, comprising two formations, with ten members which represent ten terraces of the pre-Anglian Thames. In addition, another suite of quartz and quartzite-rich sands and gravels, the deposits of a separate eastward-flowing river, can be traced from the East Midlands to the Lowestoft area (Lawson, unpublished BGS report, 1982; Clarke and Auton, 1982; Clarke, 1983; Hopson and Bridge, 1987; Rose, 1987; Bateman and Rose, 1994).

Research into various aspects of the deeper geology of East Anglia, and the basement rocks in particular, has been undertaken by a number of workers including Strahan (1913), Wills (1951, 1973, 1978), Stubblefield (1967), Linssler (1968), Owen (1971), Chroston and Sola (1975, 1982), Allsop (1985), Allsop and Jones (1981), Chroston (1985), Smith et al. (1985), Lee et al. (1990; 1991; 1993), Pharaoh et al. (1987; 1991) and Woodcock and Pharaoh (1993).

TWO

Concealed strata: Silurian to Palaeogene

The district lies towards the northern margin of a concealed platform of Palaeozoic rocks, the London–Brabant Massif (Wills, 1978). This formed a positive structural feature through Upper Palaeozoic and Mesozoic times and remained the dominant feature in the geological evolution of the region during this period.

The platform is the eroded remnant of a massif of low-grade, regionally metamorphosed, folded rocks of sedimentary and volcanic origin, together with late Caledonian intrusive rocks (Allsop, 1987; Pharaoh et al., 1987). It comprises the concealed Caledonide fold belt of eastern England and the eastern margin of the Midlands Microcraton, a subcropping terrane of Precambrian basement overlain by a thin cover, less than 3 km thick, of Cambrian to Tremadoc strata (Pharaoh et al., 1987; Smith, 1987; Lee et al., 1990; Woodcock and Pharaoh, 1993). Within the Caledonide fold belt a concealed north-west-trending tract of calcalkaline volcanic rocks lies close to the margin of the microcraton, and represents the product of arc magmatism of probable Ordovician age (Pharaoh et al., 1991).

LOWER PALAEOZOIC BASEMENT

The small number of boreholes penetrating the basement in East Anglia (Figure 3) has provided limited, but very valuable, information on the lithologies and geological history of the basement rocks (Chroston and Sola, 1982; Woodcock and Pharaoh, 1993). However, much of the research into the concealed Lower Palaeozoic and Precambrian basement has been based on geophysical studies. The early seismic refraction studies by Bullard et al. (1940) have been supplemented by more recent refraction surveys summarised by Evans and Allsop (1987), and by interpretations of the British Geological Survey's regional gravity and aeromagnetic data (Chroston and Sola, 1982; Allsop, 1984, 1985; Chroston et al., 1987). In the southern North Sea deep seismic reflection profiles (Blundell, 1993) show well-defined reflectors within the mid-crust. These have been interpreted as north-west-trending thrust structures of possible Caledonian age. Palaeogeological maps of the Lower Palaeozoic (Wills, 1978) and pre-Permian surfaces (Allsop and Jones, 1981; Smith et al., 1985) of the region have been compiled utilising both borehole and geophysical data.

These studies show the district to be underlain by Lower Palaeozoic sedimentary rocks of probable Silurian age (Smith et al., 1985), with their upper surface dipping gently to the north-east at about one-third of a degree from an estimated 250 m below OD in the south-west to nearly 500 m below OD at Lowestoft in the north-east.

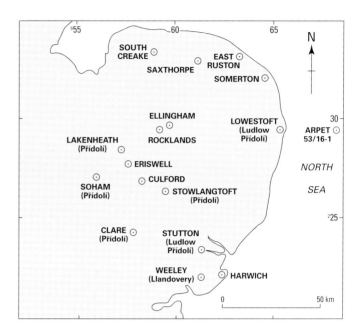

Figure 3 Distribution of boreholes proving basement in East Anglia.

Detailed examination of the available borehole cores indicates (Woodcock and Pharaoh, 1993) a systematic change in Silurian facies across East Anglia, with anoxic basin slope or outer shelf rocks possibly underlying the present district. White mica crystallinity data place the rocks in the anchizone of very low-grade regional metamorphism (Merriman et al., 1993).

The Lake Lothing Borehole at Lowestoft (Strahan, 1913), drilled in 1912 (see Appendix 2) encountered Lower Palaeozoic rocks at 492 m below OD and penetrated 62.48 m of them. Poor sample recovery precluded detailed description; however, in general the sequence comprises indurated, poorly cleaved, pale grey mudstones and shales, commonly micaceous, interbedded with lesser amounts of siltstone and sandstone. Brachiopod fragments of the genera *Lingula* and *Orbiculoidea* were assigned a general Lower Palaeozoic age by Strahan (1913); more recently these rocks have been assigned a ?Silurian age (Stubblefield, 1967; Smith et al., 1985). A core sample of cleaved mudstone from this borehole yielded a white mica crystallinity value of 0.33°2θ, indicating the low-temperature part of the anchizone.

Several boreholes have proved similar rocks just outside the district (Woodcock and Pharaoh, 1993); the nearest are at West Somerton and East Ruston (Arthurton et al., 1994), and at Rocklands (Mathers et al., 1993). The offshore borehole 53/16-1 drilled by

ARPET in 1968, 25 km east-south-east of Lowestoft, also proved Lower Palaeozoic basement (Cameron et al., 1992) of this type beneath Westphalian, Permian, Mesozoic and Palaeogene strata. Two hundred and twenty nine metres of fine-grained, well-laminated sandstone of turbiditic aspect were penetrated, with their top at 899 m below OD. The bedding lamination is disturbed by slump, convolution and bioturbation structures. The strata are unfossiliferous, and reddened beneath the basal Carboniferous unconformity. Steep bedding dips (up to 60°), the presence of weak slaty cleavage and a sonic velocity of 4.8 km per second (from well sonic log and velocity check shot survey) are typical of the Acadian (late-Caledonian) basement of East Anglia. Two core samples yield anchizonal white mica crystallinity values of 0.40°2θ, comparable with the anchizonal value obtained from the Lowestoft Borehole. The well 53/16-1 proving demonstrates the offshore continuation of the turbidite-filled Anglian Basin beneath the London–Brabant Massif of the southern North Sea.

Studies of regional gravity and aeromagnetic data provide further insights into the nature of the basement. The basic data, and the main features that can be recognised, are shown in Figure 4. The Bouguer gravity anomaly data for the district, including the offshore area (Figure 4a), show a general northward increase in values. The anomaly pattern is apparently disrupted just offshore along a north–south line, but control in this zone is poor because of lack of data. The broad Bouguer gravity anomaly maximum in the north of the district forms part of a high that extends with an east-south-easterly trend across central East Anglia and the southern North Sea into Belgium. This high has been interpreted by Lee et al. (1993) as a combined result of the presence of near-surface upper crustal Caledonian rocks and a mid-crustal magnetic body. The decrease to the south-south-east may be due partly to a concealed granite batholith in the southernmost North Sea (Lee et al. 1993, fig. 1a), but for much of the onshore areas, including the Lowestoft/Saxmundham district, it seems just as likely that the gravity decrease reflects a southward increase in the thickness of lower density Silurian and Devonian rocks. Towards the north-north-east, gravity values decrease over the low- density Mesozoic rocks of the southern North Sea (Anglo-Dutch Basin). In the crustal model shown in Figure 4d the gravity profile has been interpreted by introducing a lower-density basement (B1) which thickens southwards.

Superimposed on the regional high is a series of elongated anomalies with amplitudes of a few mGals, or less, mostly with pronounced east-south-easterly trends. Although the two sets of anomalies indicate sources of different scales, the common trend indicates that both are related to the concealed Caledonide basement. The largest of the more local anomalies are a low [centred at about 40 70] and a colinear high [centred at about 20 80] to the west. This latter anomaly, seen only in part in Figure 4a, is intersected by a north-north-east-trending high (the Sudbury–Bildeston Ridge anomaly) which is discussed below. These smaller-scale anomalies indicate

sources a few kilometres wide and 10 to 30 km long, probably at, or near, the basement surface.

These anomalies may be due to lithological changes within the basement rocks or to variations in the basement depth, both of which could be fault controlled. Of the alternatives, gravity anomalies are most easily generated by changes in the topography of the basement surface, because of the large density contrast (about 0.6 Mg/m³) between the basement rocks and the overlying Mesozoic strata (2.5 mGal for a 100 m change in depth, Figure 4d). The Sudbury–Bildeston Ridge anomaly appears to be due largely to this effect and it is improbable that the basement elsewhere is completely planar. However, buried topographic variations are not considered to be the complete explanation, for several reasons. Firstly, the Sudbury–Bildeston Ridge anomaly is characterised by a north-north-east trend, rather than the typical east-south-east trend; it may thus represent a local structural response. Secondly, anomalies similar to those in the district are found elsewhere in Britain associated with Lower Palaeozoic rocks at the surface. Thirdly, density changes could be caused by lithological variations within the basement rocks, such as the presence of low-density arenaceous facies in the Silurian and Devonian rocks, or granitic igneous bodies.

Aeromagnetic anomalies (Figure 4b) indicate that magnetic basement rocks occur at depth in the south-western part of the district, and are also present just outside the district to the west and north-west. The latter two anomalies form the eastern extremities of elongate features that cross much of East Anglia, and their extensions can possibly be traced across the district as low amplitude anomalies. Anomalies with a similar trend reappear east of the district in the North Sea, and extend into the Brabant Massif of Belgium (Lee et al., 1993), although the two magnetic terranes could be divided by a major fault trending south-south-east.

The approximately circular form of the magnetic anomaly in the south-west of the district suggests that it could be caused either by an igneous intrusion or by a fault-bounded block of magnetic basement rocks. The form and amplitude (assuming a typical magnetisation for the model) of this anomaly suggest a source at a crustal depth between 10 and 15 km.

Transient electromagnetic (TEM) soundings were made in the Thorndon area just west of the district, and a single sounding was made in the extreme south of the district near Snape [3636 5662]. The data from Snape (Figure 5) show a high resistivity layer (or layers) near surface, and an intermediate layer whose low resistivity is considered to be due to the presence of conductive saline formation water in the Chalk. A high resistivity lower layer seen at depth in the Thorndon area is indicated in the Snape data (Figure 5) only by the flattening of the curve at about 0.03 seconds onwards. This is considered to represent the top of the pre-Mesozoic basement at an interpreted depth of about 370 m below OD. The sounding provides the only determination of the depth to basement in this part of East Anglia, but the value is in agreement with earlier predictions.

a.

b.

Sheet 176
Sheet 191

c.

d.

SBR   Sudbury–Bildeston Ridge   DF   Debenham Fault
FF   Framlingham Fault   BF   Bedingfield Fault
PF   Pettaugh Fault

──┬──   Gravity lineament

──┬ M ─   Magnetic lineament,
          cross-mark on side
          with low value

──┼ M ─   Magnetic high

────   Broken lines denote uncertainty

         Magnetic
         rocks
         at depth

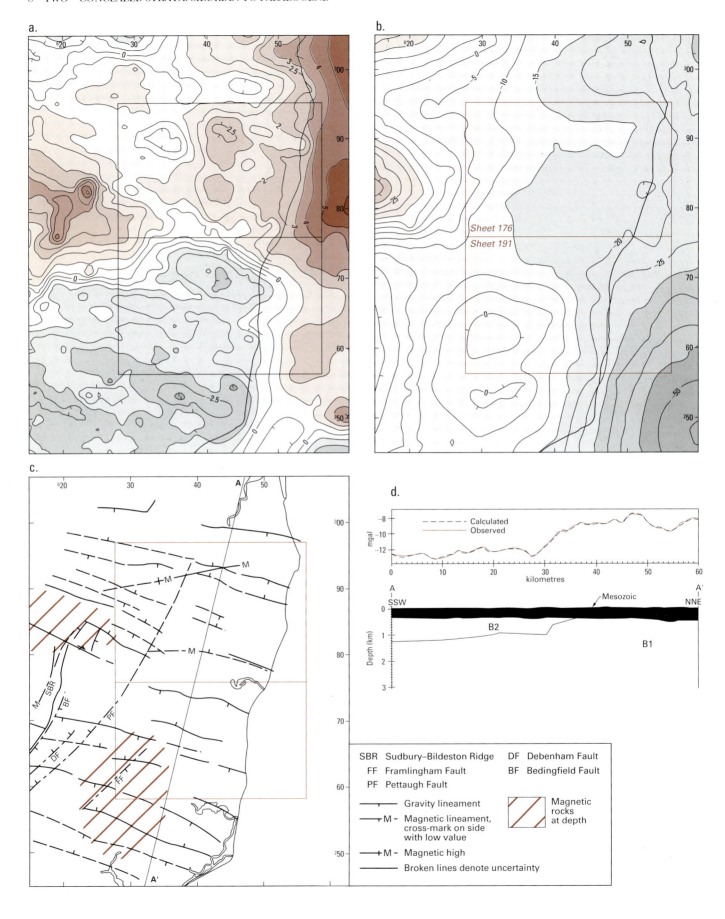

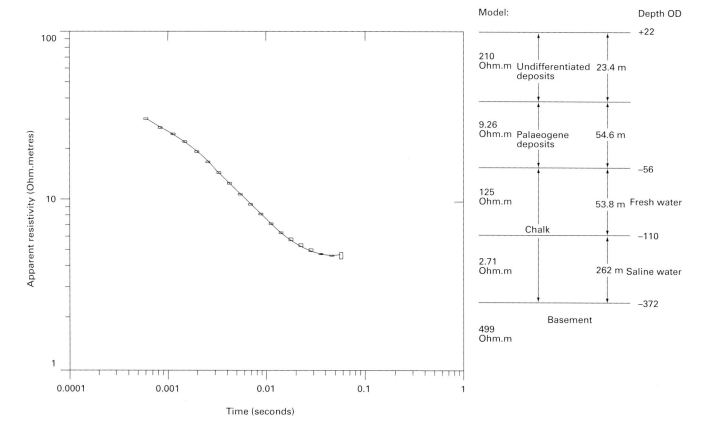

Figure 5 Transient electromagnetic (TEM) sounding for a site near Snape shown as apparent resistivity against time (points) and calculated curve (continuous line) for the model shown. Resistivities (in ohm.metre), depths (in metres) and interpreted geological sequence indicated for model layers.

MESOZOIC ROCKS

Lower Cretaceous sedimentary rocks directly overlie the Silurian basement across the entire district. To the north, Triassic and Jurassic strata are present but these thin south-eastwards and are absent from the present district. The thinning is the result of erosion and overstep by the Cretaceous strata, which in the Four Ashes Borehole [0223 7186] in the nearby Bury St Edmunds district rest directly on Siluro-Devonian basement (Bristow, 1990). In

Figure 4 Gravity and magnetic anomaly maps, main lineaments and model of crustal structure.

a. Bouguer gravity anomaly map with contours at 0.5 mGal (onshore) and 1 mGal (offshore) intervals. Density for data reduction 2.1 Mg/m³.
b. Aeromagnetic anomaly map with contours at 5 nT intervals.
c. Compilation map showing main geophysical lineaments based on gravity and aeromagnetic evidence and faults (dot-dash lines) from Bristow (1983).
d. Aeromagnetic and Bouguer gravity profile AA′ and model. Mesozoic (Density 2.10 Mg/m³), B1 – basement (2.70), B2 – low density basement (2.60). Datum level 11 mGal.

the northern half of the adjacent Great Yarmouth district Carboniferous, Permian and Triassic rocks are preserved beneath the Cretaceous strata (Arthurton et al., 1994); inland these thin southwards and die out before reaching the north of the present district (Figure 6) but offshore they have been encountered in well 53/16-1 (Cameron et al., 1992) about 25 km east-south-east of Lowestoft. Cox et al. (1989) indicated that Jurassic rocks are also present in the northern part of the Great Yarmouth district, but borehole and seismic data suggest that erosion following the Late-Cimmerian uplift has removed them from the area further south (Arthurton et al., 1994).

Cretaceous

A thick Cretaceous sequence is present at depth throughout the district (Figure 7a). The highest part, the Upper Chalk, comes to crop locally beneath drift in the northwest of the district; elsewhere it is covered by Palaeogene and Quaternary solid formations. The Cretaceous strata comprise a thin Lower Cretaceous succession of Carstone overlain by Gault and Upper Greensand, followed by the much thicker Upper Cretaceous Chalk. Within the district the only borehole penetrating the entire Cretaceous

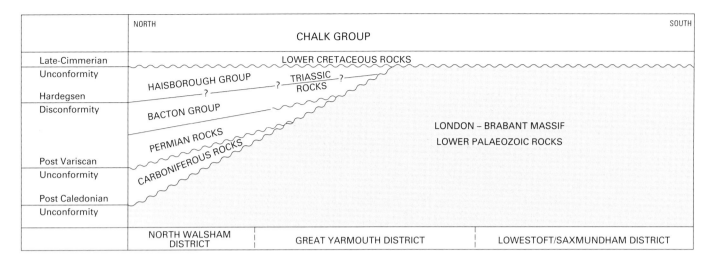

Figure 6   Schematic section showing the principal unconformities affecting the Palaeozoic and Mesozoic strata abutting the London–Brabant Massif.

sequence is at Lake Lothing, Lowestoft, but the regional disposition can be ascertained from deep boreholes in adjacent districts.

A transient electromagnetic (TEM) sounding [3636 5662] near Snape indicated the existence, above the basement, of a 262 m-thick layer with a resistivity below 3 ohm metre (Figure 5). The layer is believed to be Chalk saturated with saline water. This feature is consistent with results from soundings in the Thorndon area to the west of the district, and with observations (Bath and Edmunds, 1981) that the Chalk retains connate sea water. The overlying layer, with a higher resistivity (125 ohm m), is believed to represent Chalk saturated with fresh water. The top of the Chalk is estimated from the TEM sounding to lie at about 56 m below OD in the Snape area.

LOWER CRETACEOUS

The Carstone (Lower Greensand) proved in the Lowestoft Borehole (483.4 to 495.9 m depth) is of Albian age (Rawson et al, 1978; Gallois and Morter, 1982). It thins southwards from 16.7 m at the West Somerton Borehole [TG 4736 1935] in the North Walsham district to 12.5 m at Lowestoft. The southward thinning probably continues across the present district.

The Carstone comprises an overall coarsening upwards sequence of poorly cemented, green or white, limonitic, medium- to coarse-grained sandstone, with minor mudstone interbeds. It is inferred to have been deposited in a high energy, shallow marine shoreline environment on the north side of the London–Brabant Massif.

The Gault is the oldest preserved formation to have entirely covered the London–Brabant Massif (Owen, 1971) in Albian times (Rawson et al., 1978), overlapping earlier Lower Cretaceous formations to rest directly on Lower Palaeozoic rocks to the south (Gallois and Morter, 1982; Millward et al., 1987). The stratigraphy of the Gault in northern East Anglia has been described in detail by Gallois and Morter (1982). It thins northwards from

13.72 m in the Lowestoft Borehole (469.65 to 483.4 m depth) to 8.3 m in the West Somerton Borehole (555.7 to 564.0 m depth) and is inferred to be present beneath the entire district. The Gault comprises fining-upward cycles of pale to dark grey silty and sandy calcareous mudstones, with many erosion surfaces characterised by an abundance of phosphate- and glauconite-rich material (Gallois and Morter, 1982). It is thought to have accumulated in a shallow marine embayment between the London–Brabant Massif and a shoal to the north where Red Chalk was being deposited.

A thin development of Upper Greensand probably overlies the Gault in much of the district. In the Lowestoft Borehole 3.35 m of green clay and chalk, with black sand, was logged above the Gault between 466.34 and 469.69 m depth (Strahan, 1913) and is regarded as Upper Greensand. At Norwich, to the north-west, in the Carrow Works Borehole [TG 2414 0753] 1.83 m of sands were recorded in a similar stratigraphical position, between 349.84 and 351.67 m depth (Whitaker, 1921).

UPPER CRETACEOUS

Chalk Group

The Chalk Group underlies the entire district at depth, but is buried beneath younger solid formations except in a small area in the extreme north-west, where it is overlain by drift deposits only (Figure 2).

The Lake Lothing borehole penetrated 321.56 m (144.78 to 466.34 m depth) of Chalk. The log is insufficiently detailed to enable the thicknesses of the constituent Lower, Middle and Upper divisions to be determined, although information from nearby districts indicates that all three are likely to be present.

In the adjacent Diss district to the west the Chalk Group is about 240 m thick in the west but thickens eastwards downdip, as stratigraphically higher beds are preserved, to reach a maximum thickness of some 330 m. Of this, 45 m are Lower Chalk, 60 m are Middle Chalk and 225 m are Upper Chalk. To the north of the present

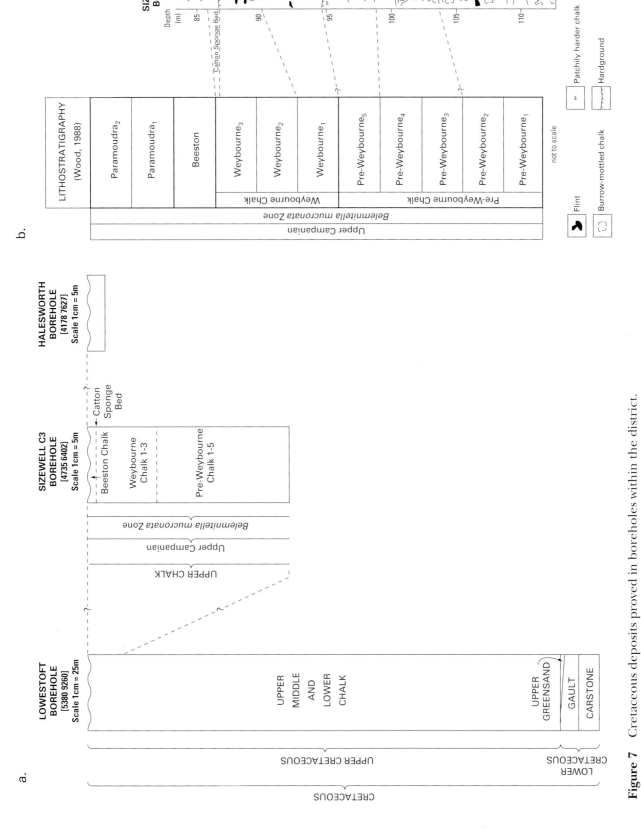

Figure 7   Cretaceous deposits proved in boreholes within the district.

a. Correlation between the Lowestoft, Sizewell C3 and Halesworth boreholes.

b. Correlation of the Sizewell C3 Borehole with the lithological subdivisions of the *Belemnitella mucronata* Zone Chalk of Norwich.

district, in the Great Yarmouth area, the Chalk reaches a maximum thickness of about 460 m, of which 30 m are Lower Chalk, 60 m are Middle Chalk, and 370 m are Upper Chalk. The full potential thickness of Upper Chalk is greater still, but the top has been eroded throughout East Anglia.

Chalk is a very pure micritic limestone composed mainly of the disaggregated skeletons of planktonic coccolithophorid algae. The skeletal material usually occurs as individual laths and plates, a few microns across, and more exceptionally as rings of laths or coccoliths. Macrofossils occur throughout, but tend to be concentrated in distinct layers. The main faunal groups represented are echinoids, crinoids, belemnites, ammonites, bivalves, gastropods, brachiopods and sponges.

The local lithostratigraphy and biostratigraphy of the Chalk beneath the Lowestoft/Saxmundham district is poorly known, because of the absence of outcrop and paucity of borehole data. The sequence can be extrapolated from boreholes and sections in nearby districts, although in the adjacent Great Yarmouth district to the north direct information is also scarce.

The biostratigraphy, lithostratigraphy and structure of the Chalk in Norfolk have been reviewed and revised by Peake and Hancock (1961, 1970). The lithostratigraphy, biostratigraphy and history of research of the Chalk in the Norwich district were further updated by Wood (1988), from studies carried out as part of the geological survey for the Norwich map (Sheet 161). A summary of Wood's work appears in the Norwich memoir (Cox et al., 1989).

On the evidence from boreholes outside the district the Lower Chalk is less pure than the Middle and Upper Chalk. It is typically free of flints, is grey in colour, and includes some hardgrounds and several horizons of dark grey, shell-detrital silty chalk. The Middle Chalk is whiter than the Lower Chalk, is almost flint-free (two thin bands have been recorded in the Trunch Borehole), and has marl seams. Beds of shell-detrital chalk are common in the lower part. The Upper Chalk is white, although it has been described locally at outcrop as yellow. It contains many courses of tabular and nodular flints.

A preliminary account of the biostratigraphy of the Trunch Borehole [2933 3455] sunk in the Mundesley district of north Norfolk in 1975 was given by Gallois and Morter (1976). A revised account based partly on unpublished logs by Mr A Morter can be found in the Great Yarmouth Memoir (Wood et al., in Arthurton et al., 1994, appendix 1). This borehole proved a total Chalk thickness of 466.52 m and represents the most complete cored sequence of the Chalk Group in Britain. The chronostratigraphical range of the Chalk in the borehole is from Maastrichtian to the base of the Cenomanian. The Middle Chalk is condensed in comparison with that of the Northern Province, with which it has the greatest affinity (Wood et al., in Arthurton et al., 1994, appendix 1). The Lower Chalk is similarly comparable to that in the Northern Province, but is very condensed.

The BGS Halesworth (Heath Farm) Borehole [4178 7627] (Figure 7a), drilled during the present survey, cored 1.89 m of Upper Chalk from the depth interval 38.12 to 40.17 m below the surface. The chalk is massive,

soft, white, and lacks any distinct marl or flint horizons. The fauna includes the bivalves *Neithea sexcostata*, *Pseudoptera caerulescens?*, oyster and inoceramid fragments; the brachiopods *Carneithyris?*, *Cretirhynchia?*, *Kingena* and an unidentifiable terebratulid; the belemnite *Belemnitella* sp.; plus bryozoan?, scaphopod?, cidaroid spine and fish fragments (Woods, 1993a; 1994). Elsewhere, the brachiopod *Carneithyris* first appears in the 'Weybourne₃' Chalk and then ranges into the overlying Beeston Chalk (Wood, 1988). The specimen of *Belemnitella* from near the base of the sequence (40.08 m) is of uncertain specific affinity, but the genus commonly occurs in the upper *Gonioteuthis quadrata* and succeeding *B. mucronata* zones (Christensen, 1991). The presence of common inoceramid shell fragments in the core led Woods (1994) to suggest that the strata might be equivalent to the chalk above the Catton Sponge Bed (Beeston Chalk) of the Norwich area, as this is characterised by 'apparently laterally continuous belts of chalk rich in large fragments of *Inoceramus* shell' (Wood, 1988).

During the survey the Upper Chalk was also examined from the Sizewell C3 Borehole [4735 6402] (Figure 7b) which encountered Upper Chalk beneath Palaeogene strata from 83.06 m depth to the base of the borehole at 112.65 m. The chalk in the borehole is of Upper Campanian age, and restricted to the *Belemnitella mucronata* Zone. The sequence (Figure 7b) has been correlated (Woods, 1993b) with the Chalk of the Norwich area (Wood, 1988), though with some uncertainties. A full discussion of these is given by Woods (1993b).

The lowermost chalk in the borehole (98.82 to 112.65 m depth) probably correlates with the Pre-Weybourne₁₋₄ Chalk of the Norwich area. In the borehole this interval is a soft white chalk with marly intervals and mottled horizons. The mottled texture is produced by diffuse concentrations of more or less marly chalk, and may be the result of bioturbation. Occasional flints are present, including a fine burrow-form example at 111.7 m. The fauna includes brachiopod and bivalve species (Woods, 1993b) characteristic of the Upper Campanian *Belemnitella mucronata* Zone, notably *Kingena pentangulata*, *Gyropleura inequirostrata* and *Pseudolimea granulata* (Owen, 1970; Cleevely and Morris, 1987; Woods, 1908).

The chalk between 97.02 m and 98.82 m is divided by a prominent flint at 97.75 m. The lower interval is a smooth soft white chalk, lacking marl. Immediately below the flint the chalk becomes marlier and bioturbated; this lithology continues to the top of the interval. The fauna is dominated by brachiopods, referred to *Cretirhynchia* ex gr. *lentiformis*, and echinoid spine and test fragments.

The lower part of the interval between 94.75 m and 97.02 m is characterised by smooth white chalk, lacking marl. At 96.15 m is a thin marly horizon, above which there is an upward passage, via marly burrowed chalk, into indurated chalk capped by an irregular glauconitised hardground. Compared to the underlying interval, *C.* ex gr. *lentiformis* is much less frequent, and there is a significant increase in the oyster fauna, especially the sudden appearance of *Hyotissa? semiplana*. The hardground is associated with common sponge remains, some of which are phosphatised. *Belemnitella mucronata* is also present; this first appears in the Pre-Weybourne$_5$ Chalk of Norwich (Wood, 1988). The simplest correlation with Norwich is to relate this unit, and the underlying interval (described above) with Pre-Weybourne$_5$ and/or Weybourne$_1$.

The interval from 90.15 m to 94.75 m contains soft, bioturbated chalk with frequent marly horizons. A major flint occurs at 90.5 m, underlain by a core loss of about 0.75 m, below which are scattered occurrences of small and fragmentary flints. Glauconitic phosphate clasts near the base of the unit are related to the underlying hardground, but glauconitic chalk clasts at higher levels do not appear to be associated with bed induration. The fauna is characterised by frequent specimens of *C. ex gr. lentiformis*. Echinoid fragments are common in the lower part but become less frequent above 93.75 m. The top of the interval approximates to the upper limit of *C. ex gr. lentiformis* and *C. lentiformis?*, which in Norfolk do not range above Weybourne$_1$ (Wood, 1988).

The top of the interval between 85.82 m and 90.15 m is characterised by an irregular glauconitised hardground, associated with a large glauconitised pebble and a worn belemnite guard fragment. Sponges and gastropods also feature. The indurated chalk beneath this surface contains glauconitised phosphatic clasts and more diffuse patches of pale buff phosphatic chalk. A further glauconitised hardground is present about 0.75 m below the top of the unit. Below this the chalk becomes softer and mottled, with frequent impersistent marly horizons, and strongly defined subhorizontal trace fossils, perhaps *Zoophycos*, from about 87.6 m to 87.8 m. The remainder of the interval is soft white mottled chalk with marly chalk infilling burrows. The base of the unit is a marly chalk horizon. The presence of *Pseudolimea granulata* within this interval in the borehole suggests the *B. mucronata* Zone, and *Cretirhynchia woodwardi* (found at about 88.95 m in the borehole) does not range above the Catton Sponge Bed in the Norwich Chalk (Wood, 1988).

The highest chalk present in the borehole (83.06 m to 85.82 m depth) has a rather massive uniform appearance; there is no obvious bioturbation, and the only significant marly horizon, at 85.65 m, may be due to the injection of drilling mud. The lower part of the interval contains patches of very hard, locally iron-stained chalk. The highest chalk is massive, soft and white with scattered pale phosphatic and glauconitised clasts. Phosphatised fossils occur throughout the interval, but especially towards the base. Two nodular flints were recorded near the middle of the unit. The hardground that underlies this interval appears to mark a major increase in macrofaunal abundance and diversity. Frequent sponge remains characterise the whole of this interval, but are especially common in the lower part, where gastropods are also concentrated. The ammonites *Baculites* sp., *?Nostoceras* (*Bostrycoceras*) and *?Nostoceras* are also present. The records of *Cretirhynchia arcuata* and *Carneithyris?* are crucial to the interpretation of this interval. In Norfolk, both of these taxa first occur near the top of Weybourne$_3$, close to the Catton Sponge Bed, and range into the overlying Beeston Chalk (Wood, 1988).

Although not part of the Catton Sponge Bed as here interpreted, the rich sponge fauna that is especially common near the base of the interval, is possible further evidence for correlating the underlying hardground with the terminal Catton Sponge Bed horizon.

In many borehole logs the uppermost few metres of the Chalk are recorded as 'putty chalk'. As this lithology occurs at varying stratigraphical levels across the district it would appear to be a weathering phenomenon rather than an original feature of the Chalk. However, in some instances the soft texture may have been caused by flint fragments from the overlying Bullhead Bed remaining down the hole during drilling, and grinding the underlying chalk into a flour. This would also account for the unusually thick Bullhead Bed commonly recorded in borehole logs.

PALAEOGENE ROCKS

Palaeogene strata are confined to the eastern half of the district (Figure 8), and are concealed by Quaternary marine and glacial deposits. They rest unconformably on the Upper Chalk, dipping very gently eastwards at about 1°. The strata vary considerably in thickness up to a

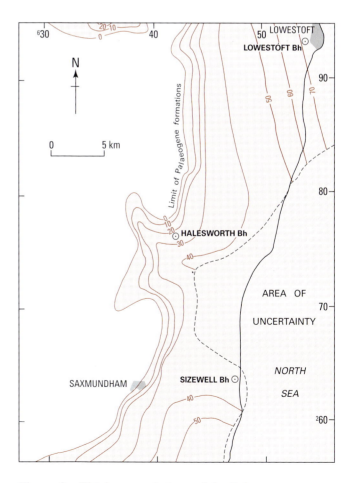

Figure 8 Thickness variations of the Palaeogene formations: isopachytes in metres.

maximum recorded 72 m in a borehole near Lowestoft (Figure 8). Their western limit has the form of a low, buried escarpment, a feature noted also in the adjacent Great Yarmouth district (Arthurton et al., 1994).

By early Palaeogene times, a geography similar to that of the present day had evolved, and an emergent British Isles lay to the west of a small epicontinental sea (Figure 9). After several phases of uplift in the late Cretaceous/ early Palaeocene, during which Cretaceous and Palaeocene chalks were eroded (Knox, in press; Neal, in press), prolonged subsidence began along the central rift zone of the North Sea Basin. However, subsidence in the south of the basin was relatively slow and punctuated by short phases of uplift. Eustatic changes in sea level, superimposed on this tectonic pattern, produced the transgressive/regressive cycles and the interplay of paralic to open marine shelf depositional environments characteristic of these deposits.

Numerous ash layers in the sediments chronicle the intense volcanic activity associated with contemporaneous rifting of the north Atlantic (Knox, 1984). Several authors have related North Sea Basin tectonism to the crustal stress pattern of the North Atlantic (e.g. Galloway et al.,1993). Knox (in press), suggests precise correlations between the phases of volcanic activity and episodes of uplift in the North Sea Basin.

NOMENCLATURE

The primary lithostratigraphical framework of Palaeogene deposits in the south and east of England was established in the mid 19th century by Joseph Prestwich. His nomenclature, presented in a series of publications (1847, 1850, 1852 and 1854) was, with minor amendments (Table 1) by Whitaker (1866, 1889) in use for over a hundred years. Notwithstanding further revision by King (1981), it remains the basis of the most recent classification (Table 2), proposed by Ellison et al. (1994).

PALAEOCENE–EOCENE BOUNDARY

A detailed biostratigraphy has been established for the Palaeogene from oceanic sequences, using foraminiferids, calcareous nannoplankton and dinoflagellates (Table 3). However, the location of the Palaeocene/ Eocene boundary in the deposits of north-western Europe has proved elusive. The difficulty comes partly from a lack of international agreement on the position of the boundary (see for example King, 1981), and partly from the paucity of diagnostic marine fauna and flora. The boundary is currently taken at the base of calcareous nannoplankton Zone NP10 of Martini (1971), which corresponds to a depositional hiatus in the late Palaeocene deposits of north-west Europe.

A marked change in the isotopic composition of carbon (a shift of $\delta^{13}C$ to lighter values) has recently been recorded in oceanic Palaeogene strata, within Zone NP9 and very close to the NP9–NP10 boundary. It occurs in Chron 24R of the magnetostratigraphic notation (see

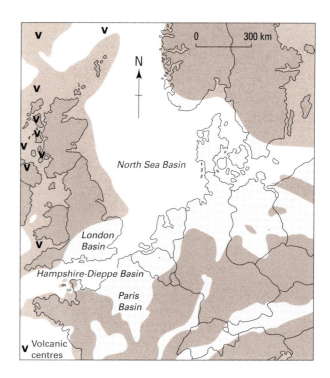

Figure 9   Generalised palaeogeography of north-west Europe during the early Palaeogene.
Simplified after Ziegler (1982) and Cameron et al. (1992).

v   volcanic rocks

p.17) at a numerical age of approximately 55.3 Ma (Cande and Kent, 1992; Thomas and Shackleton, 1993). This change in isotopic concentrations corresponds to a global extinction of benthic foraminifera, a major event in the Earth's history. Sinha and Stott (1993) have suggested that carbon isotope stratigraphy may offer the best means of correlating marine and terrestrial sequences because their carbon pools are linked by the atmosphere, citing as an example their identification of the $\delta^{13}C$ shift in terrestrial Sparnacian (late Palaeocene) deposits of the Paris Basin. Thus, the Palaeogene $\delta^{13}C$ shift may, in future, be adopted to define the Palaeocene–Eocene boundary.

Ormesby Clay Formation

The Ormesby A Borehole [5145 1425] (Figure 10), drilled by BGS in 1982 (Cox et al., 1985), encountered a sequence of mudstones at the base of the Palaeogene. Knox et al. (1990) introduced the informal term Ormesby Clay for these mudstones, and subsequent borehole provings throughout East Anglia have established its regional significance. The name was later formalised as Ormesby Clay Formation by Ellison et al. (1994).

A positive area, the Ipswich–Felixtowe High (Ellison et al., 1994), appears to have had a significant effect on the distribution of facies in East Anglia and the London Basin. It separated the depositional area of the Ormesby Clay Formation from that of the coeval Thanet Sand Formation, and its influence probably persisted into the earliest Eocene (Figure 11).

Table 1 Main developments in the classification of the early Palaeogene of south-eastern England (modified after King, 1981).

Prestwich (1847–54)	Whitaker (1866)	Whitaker (1889)	King (1981)			
London Clay	London Clay	London Clay	Walton Member		A2	London Clay Formation (part)
			Harwich Member	Swanscombe Member	A1	
Basement Bed	Basement Bed	Basement Bed	Tilehurst Member			Oldhaven Formation
	Oldhaven Beds	Oldhaven Beds	Herne Bay Member			
		Blackheath Beds	Blackheath Beds			
Woolwich and Reading Series	Woolwich and Reading Beds	Woolwich and Reading Beds	Woolwich and Reading Beds			
			'Bottom Bed' (Hester, 1965)			
Thanet Sands	Thanet Beds	Thanet Beds	Thanet Formation (Ward, 1978)			

LITHOSTRATIGRAPHY AND SEDIMENTOLOGY

The formation comprises variably glauconitic mudstones up to 27 m thick, with a basal flint gravel (Bullhead Bed) that rests unconformably on the Upper Chalk. This basal gravel is commonly less than 20 cm thick and consists of unworn glauconite-coated flints, up to cobble grade, in a glauconitic sand matrix. Records of greater thickness, up to 2 m, are probably unreliable.

Knox et al. (1990) recognised four mudstone units, OC1-4, in the formation in Norfolk. The oldest of these, unit OC1, is an olive-grey poorly bedded silty mudstone, variably glauconitic, with three thin altered volcanic ash layers. The ash layers belong to Phase 1 of the North Atlantic early Palaeogene pyroclastic activity (Knox and Morton, 1988). The top of the the unit is marked by burrows filled with reddish brown clay derived from the base of the overlying unit. Unit OC2 comprises pale reddish brown mudstone which passes up into pale grey, bioturbated, slightly calcareous and glauconitic mudstone. The latter is sharply overlain by unit OC3, a greyish brown, poorly bedded, calcareous, strongly glauconitic mudstone that becomes siltier and less glauconitic upwards. The succeeding unit OC4 comprises a dark greyish brown waxy mudstone, with an altered volcanic ash layer close to the base.

The potential of the reddish brown mudstone of unit OC2 as a regional marker horizon was recognised by Knox et al. (1990) and developed by Jolley (1992). He noted the occurrence of this mudstone at many localities in East Anglia and offshore of Norfolk, and at the Thanetian type locality at Pegwell Bay in Kent. At most places it is at, or near, the base of the Ormesby Formation, but in Norfolk the mudstone overlies several metres of silty mudstone of unit OC1. For example, in the Hales Borehole [3671 9687], located less than 2 km north of the district, 4 m of reddish brown mudstone were recorded above 10 m of unit OC1, whilst 21 km to the south in the Halesworth Borehole [4178 7627] unit OC1 was absent and 0.9 m of the reddish brown mudstone lay directly on Upper Chalk.

All four units of the formation were recorded in the Hales Borehole but only units OC2 and OC3 were present in the Halesworth Borehole and in the Sizewell C3 Borehole [4735 6402] in the south-east of the district (Figure 10). This indicates southward onlap, and later erosion of unit OC4 during regression.

BIOSTRATIGRAPHY

The age of the formation, as dated from dinoflagellate cysts, ranges between the *Palaeoperidinium pyrophorum* and *Alisocysta margarita* zones of Powell (1992), which themselves are within the range of nannoplankton zones NP6 to NP8 (Table 3). The base of the formation is diachronous, increasing in age northwards.

Table 2 Lithostratigraphical classification and correlation of early Palaeogene strata of the London Basin, East Anglia and the Southern North Sea Basin.

LITHOSTRATIGRAPHY OF ELLISON ET AL. (1994)			UNITS OF ELLISON ET AL. (1994) RECOGNISED IN THIS DISTRICT		NORTH SEA BASIN FORMATIONS
GROUP	FORMATION	INFORMAL AND FORMER USE			
THAMES GROUP	LONDON CLAY FORMATION		?LONDON CLAY FORMATION (Walton Member)		BALDER FORMATION
	HARWICH FORMATION	'SWANSCOMBE MEMBER' 'TILEHURST MEMBER' 'HARWICH MEMBER' OLDHAVEN BEDS BLACKHEATH BEDS 'HALES CLAY'	'HARWICH MEMBER'	HARWICH FORMATION	
			'HALES CLAY'		SELE FORMATION
LAMBETH GROUP	READING FORMATION / WOOLWICH FORMATION	'READING BEDS' / 'WOOLWICH BEDS'	LAMBETH GROUP		SELE FORMATION
	UPNOR FORMATION	'BOTTOM BED'			
	THANET SAND FORMATION / ORMESBY CLAY FORMATION	'THANET SANDS' 'THANET BED' THANET FORMATION / 'ORMESBY CLAY'	ORMESBY CLAY FORMATION		LISTA FORMATION

MAGNETOSTRATIGRAPHY

The polarity of the geomagnetic field has reversed many times throughout the Earth's history. The 'imprint' of the prevailing geomagnetic field is retained in rocks at the time of their formation. This phenomenon has given rise to a stratigraphical method based on the magnetic polarity record of the rocks. By convention the polarity is said to be *normal* (n) when the field is orientated towards the geographic north, as at the present, and *reversed* (r) when it is orientated to the south. A *chron* is the term used to describe a main subdivision of time based on polarity.

The Ormesby Clay Formation in the Ormesby and Hales boreholes ranges from Chron C26r, the oldest onshore Palaeocene, to C25r. Of particular importance is the coincidence of the reddish brown mudstone in unit OC2 with a period of normal polarity, Chron 26n (Knox et al., 1990). The basal, reddish brown, deposits in the Thanetian stratotype also coincide with Chron C26n and have been assigned by Siesser et al. (1987) to zones NP6/NP7. From this association, and the regional persistence of the reddish brown horizon, noted above, the restriction of the oldest onshore Palaeocene deposits (Chron C26r) to the northernmost part of the district and Norfolk has been determined.

LAMBETH GROUP

Ellison et al. (1994) introduced this term to replace the 'Woolwich and Reading Beds' of Whitaker (1866). The group includes the Upnor Formation (formerly the 'Bottom Bed'), the Woolwich Formation and the Reading Formation, and is equivalent to the lower part of the off-shore Sele Formation. Deposits of the appropriate age and of similar, but not identical, lithologies were proved in boreholes at Sizewell in the south of the district and at Halesworth in the centre of the district. They were absent from the boreholes at Hales and Ormesby north of the district (Knox et al., 1990), but the log of the Lowestoft (Lake Lothing) Borehole [5380 9260], drilled in 1912 to provide water for the East Anglian Ice Company, recorded deposits of bluish or reddish mottled clay, with layers of sand and buff 'earth' between 121.9 and 136.4 m depth. This suggests the possible presence of deposits representing the Lambeth Group in the extreme north-east of the district. It seems probable, therefore, that the Lambeth Group extends throughout most of the Palaeogene subcrop within the district; the maximum proved thickness in a borehole is 12.65 m, but as thickness data are only available for five boreholes in an area of several hundred square kilometres, this may be unrepresentative.

Table 3 Application of the magnetic and biostratigraphical record to the Palaeogene deposits of south-east England. (Calibration of magnetostratigraphy and nannoplankton zones after written communication from W A Berggren to R W O'B Knox, August 1994).

MAGNETOSTRATIGRAPHY (Conde and Kent, 1992)			CALCAREOUS NANNOPLANKTON ZONES (Martini, 1971)	DINOFLAGELLATE CYST ZONES		CARBON ISOTOPE EVENT	LITHOSTRATIGRAPHY (Ellison et al., 1994)	STAGE	EPOCH
TIME million years before present	CHRONS	POLARITY n, normal r, reversed		(Costa and Manum, 1988)	(Powell, 1992)				
54 —	C24	r (n)	NP11				LONDON CLAY FORMATION	YPRESIAN	EOCENE / EARLY
			NP10	D6	Wme/Was (W. meckfeldensis/ W. astra)				
55 —				D5b	Gor (G. ordinata)		HARWICH FORMATION	?	
56 —	C25	r (n)	NP9	D5a	Aau (A. augustum)	δ¹³C ←	WOOLWICH and READING FORMATIONS	THANETIAN	LATE PALAEOCENE
					Ahy (A. hyperacanthum)		UPNOR FORMATION		
57 —			NP8	D4	Ama (A. margarita)		ORMESBY CLAY FORMATION / THANET SAND FORMATION		
			NP7						
58 —	C26	r (n)	NP6		Ppy (P. pyrophorum)			SELANDIAN	
59 —			NP5	D3	Csp (C. speciosum)				
			NP4						

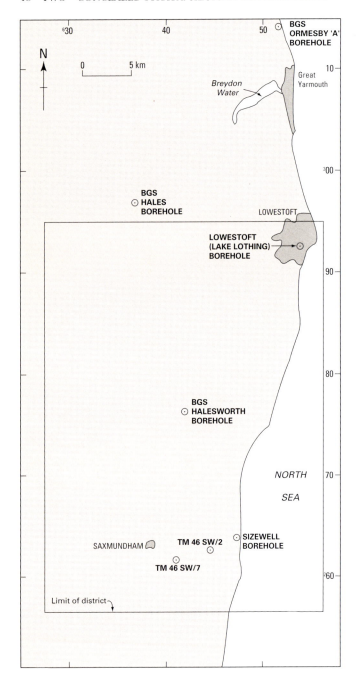

Figure 10   Location of important boreholes proving Palaeogene deposits in the district and adjacent areas.

Detailed lithological descriptions are available from two boreholes and broad descriptions from two others.

In the Halesworth Borehole (Appendix 2), 7.34 m of strata have been assigned to the Lambeth Group. At the base, 4.29 m of lenticularly interbedded to finely interlaminated dusky yellowish brown silty clay and pale yellowish brown medium-grained sand, rest sharply on the Ormesby Clay Formation. The base is probably erosional; offshore seismic profiles show that the Lambeth Group/Sele Formation deposits are commonly incised

into the Ormesby Clay/Lista Formation (Cameron et al., 1992). Locally, Lambeth Group sand fills well-defined burrows into the Ormesby Clay (Plate 2a). The lenticularly bedded strata pass up into 0.45 m of shelly sand and thin lignitic layers alternating with greyish brown clay, which in turn pass up into 1.45 m of thinly-bedded yellowish brown sands and grey clays (Plate 2b). These are overlain by 1.15 m of light bluish grey and yellowish brown mottled clay with rootlet traces (Plate 2c). The Sizewell C3 Borehole [4735 6402] showed a similar basal sequence but this was overlain by several metres of yellowish brown medium-grained sand with sporadic clay layers, followed by reddish brown silty clays. This pattern is mirrored in two less rigorously described boreholes in the south-east of the district; these are borehole TM46SW/7 [4095 6170] at Moor Farm, Friston where 9 m of clay were recorded, and borehole TM46SW/2 [4439 6260] at Leiston Ironworks where 8 m of sandy deposits overlie clays and 'loam'.

The Upnor Formation has not been proved in the district and is thought to be absent. The lenticular-bedded sands and clays at the base of the sequence are considered to be intertidal and lagoonal deposits equivalent to the Woolwich Formation, though of slightly different lithologies. The overlying sands occupy fluvial incisions into the clays and are equated with the Reading Formation, whilst the mottled clays with rootlet traces are typical of pedogenic horizons within that formation.

No useful biostratigraphical data are available for these deposits within the district, as they have proved barren of calcareous nannofossils and palynomorphs. However, their stratigraphical relationship with the Ormesby Clay and Harwich formations indicates a late Palaeocene age within nannoplankton zone NP9.

THAMES GROUP

Harwich Formation

The term Harwich Formation has been extended by Ellison et al. (1994) to encompass all the deposits between the Lambeth Group and the London Clay (Table 2) throughout the London Basin and East Anglia. In the present district this stratigraphical interval is occupied by a 'distal' facies of the formation referred to here (following Ellison et al., 1994) as the Hales Clay Member and the 'Harwich Member'. These units occur in the Ormesby and Hales boreholes north of the district, and in the Sizewell and Halesworth boreholes within the district. Offshore equivalents are the upper part of the Balder Formation and the lower part of the Sele Formation (Knox and Holloway, 1992).

Hales Clay Member

Knox et al. (1990) recognised two units with a combined thickness of 15 m in the Hales Borehole: a pale brown siltstone with rare thin sand layers, and overlying greyish

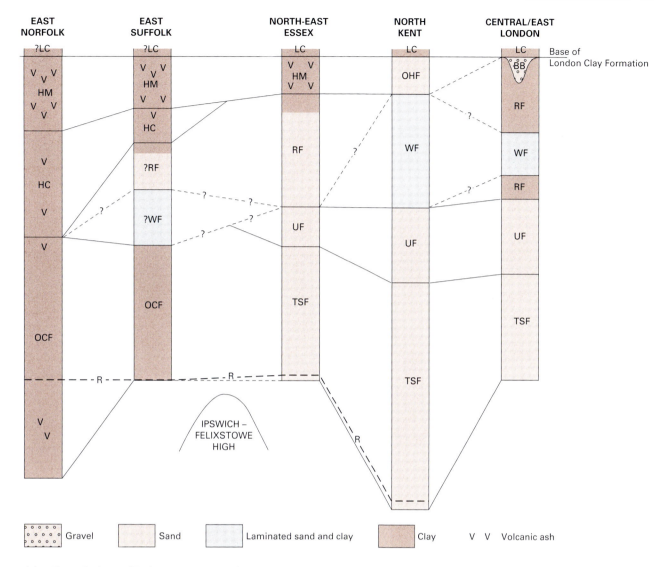

EAST NORFOLK	EAST SUFFOLK	NORTH-EAST ESSEX	NORTH KENT	CENTRAL/EAST LONDON

Base of London Clay Formation

Gravel   Sand   Laminated sand and clay   Clay   V V Volcanic ash

Figure 11   Correlation of Palaeogene strata between East Anglia and Kent.

BB = Blackheath Beds, HC = Hales Clay Member, HM = 'Harwich Member'; LC = London Clay Formation,
OCF = Ormesby Clay Formation, OHF = Oldhaven Formation, R = 'Reddish brown mudstone base',
RF = Reading Formation, TSF = Thanet Sand Formation, UF = Upnor Formation,
WF = Woolwich Formation.

brown silty and sandy mudstones. Four bentonised volcanic ash layers belonging to Subphase 2a of Knox and Morton (1983), were also recorded compared to ten in a borehole 50 km offshore (Morton and Knox, 1990). It has not been possible to subdivide the Hales Clay in the present district, though about 4 m of pale brown burrowed siltstone with sporadic thin sand layers, similar to the basal unit at Hales, were encountered in the Halesworth Borehole. No ash layers were identified.

The basal unit at Hales contained dinoflagellate assemblages of nearshore aspect, comprising a few non-diagnostic species with much wood and cuticle material (Knox et al., 1990). The Hales Clay as a whole probably accumulated in shallow marine environments. Knox et al. (1990) regarded them as an early phase of the southward transgression of the 'London Clay sea'.

'Harwich Member'

The 'Harwich Member' comprises olive-grey to greyish brown sandy siltstones, sporadically glauconitic, with numerous basaltic ash layers (Knox and Ellison, 1979; King, 1981); 86 in the Ormesby Borehole and reportedly over 300 in the central North Sea Basin (Malm et al., 1984). These ash layers are generally less than 50 mm thick, are reworked and bioturbated, and belong to the main Palaeogene episode of pyroclastic activity, Subphase 2b of Knox and Morton (1988). The Harwich Stone Band, a more or less continuous layer of tabular calcareous concretions near the base, has been identified in boreholes and is a prominent reflector in offshore seismic profiles. Thickness data for the 'Harwich Member' in the district are too sparse to be useful but it has a maximum recorded thickness in East Anglia of

a.

b.

c.

Plate 2   Lambeth Group lithologies in the Halesworth Borehole (GS 561).

a. Sand-filled burrows in clay.

b. Thinly bedded sands and clays.

c. Mottled clay with rootlets (top of Lambeth Group) overlain by Hales Clay Member.

Scale   core is approximately 10 cm wide

about 26 m in the Ormesby Borehole and about 16 m to the south of the district. The depositional environment was shallow marine, near to offshore open shelf.

BIOSTRATIGRAPHY

The Harwich Formation has not yielded any indigenous calcareous microfauna (Ellison et al., 1994), but the dinoflagellate cyst assemblages, including *Apectodinium* species, indicate dinoflagellate Zone D5, possibly Subzone 5b, of Costa and Manum (1988; Harland, 1993). Many calibrations of the different zonal schemes have been attempted and are constantly revised. Costa and Manum (1988) calibrated their Zone D5 to nannoplankton Zone NP9. On this basis the Harwich Formation was assigned to the late Palaeocene, a practice followed by BGS offshore surveys (Cameron et al., 1992). However, correlation of the ash layers with those of the north-east Atlantic by Knox (1984) places the formation in nanno-plankton Zone NP10. Powell (1992) calibrated his dinoflagellate *G. ordinata* Zone to Subzone D5b of Costa and Manum (1988) and to the lower part of NP10. This places the formation in the early Eocene, an assignation adopted by Ellison et al. (1994) and followed here.

London Clay Formation

Little information on the nature and distribution of this formation exists for the district. It was not proved in the Hales and Halesworth boreholes, or in the Sizewell C3 Borehole, having presumably been removed by post-Palaeogene erosion. However, it may be present locally where thick sequences of 'brown' or 'blue' clays are recorded in the logs of old wells and boreholes, and it may occur offshore. South of the district in north-eastern Essex, King (1981) assigned 'silty clays and clayey silts, with sand partings and sand laminae at some levels and occasional beds of sandy silt' overlying the 'Harwich Member' to his Division A2 (Walton Member) of the London Clay Formation. The base of the London Clay Formation, as redefined by Ellison et al. (1994), is marked by a marine transgression although, from King's account, the upward change in lithology from 'Harwich Member' to Walton Member is subtle.

DEPOSITIONAL SEQUENCE OF PALAEOGENE DEPOSITS

Many authors have applied sequence stratigraphical analysis to the Palaeogene deposits of the North Sea Basin. Perhaps the most comprehensive account is that of Neal (in press), in which five major regression/transgression facies (R/TF) cycles are described and correlated.

Knox (1996) has recognised three major sequence boundaries in Palaeocene deposits of south-eastern England. These define partial second-order sequences, or 'sequence sets' as Knox terms them, which are equivalent to the lithostratigraphic divisions of Prestwich (see Table 1), and are closely related to the three oldest R/TF cycles of Neal (1996). The sequences are incomplete, lacking lowstand facies that were deposited in the deeper water of the central North Sea Basin (see, for example Neal (1996)), because they accumulated at the southern margin of the basin.

The base of Knox's (1996) Ormesby/Thanet Sequence Set is marked by an unconformity with the Upper Chalk. In East Anglia the lowest units, OC1-3, of the Ormesby Clay Formation represent successive phases of a late Palaeocene marine transgression. The regressive phase of the set may be represented by unit OC4.

The lower boundary of the succeeding Lambeth Sequence Set is defined by an unconformity at the base of the Upnor Formation. This formation, which is trans-gressive in character, is absent from the district and the Woolwich Formation rests directly on unit OC3 of the

Ormesby Clay, implying that an unconformity exists at its base. The Woolwich and Reading Formations are high-stand and aggradational deposits that accumulated in intertidal to fluvial environments.

At the end of the cycle, a phase of uplift in marginal areas of the southern North Sea Basin caused widespread erosion of Lambeth Group deposits. In Norfolk they have been completely removed so that the Hales Clay Member rests directly on the Ormesby Clay. This unconformity marks the lower boundary of the Thames Sequence Set. According to Knox (in press), the transgressive Harwich Formation represents the return of marine conditions in south-eastern England, following the collapse of the Southern North Sea Basin uplift of Lambeth Group times.

MAGNETIC SUSCEPTIBILITY

Magnetic susceptibility is a measure of the ease with which a body becomes temporarily magnetised by a magnetic field. In a sediment it is controlled by the concentration and composition of the magnetisable minerals present. Primarily it depends upon the concentration of ferrimagnetic minerals (iron-titanium oxides), of which magnetite (Fe_3O_4) is the most important because of its common occurrence and high susceptibility values. Secondary contributions to the total magnetic susceptibility come from paramagnetic substances, which are materials with small positive values of magnetic susceptibility. These include clay minerals, particularly chlorite, smectite and glauconite, ferromagnesian silicates, iron and manganese carbonates, iron disulphides (for example pyrite), and authigenic ferric-oxyhydroxide mineraloids collectively referred to as limonite. Weathering generally reduces magnetic susceptibility because magnetite is oxidised to haematite which has a much lower susceptibility.

Apparent magnetic susceptiblity was measured on cored material recovered from the Hales, Halesworth, and Sizewell C3 boreholes, at intervals initially of 10 cm and subsequently of 5 cm. The measurements were carried out using a hand-held KT-5 micro-kappameter (manufactured by Geofyzika Brno). The kappameter consists of a 10 kHz oscillator and an induction coil within a circular measuring face of 60mm diameter. The frequency of the oscillator is measured in free space and then against the surface of the core. Apparent susceptibility is computed from the frequency difference and displayed in units of 1×10^{-3} SI. Various corrections can be applied to obtain the 'true' susceptibility, which depends on the size and shape of the rock object measured. However, as only relative changes were required for correlation purposes, the 'apparent' susceptibility was appropriate.

The variation in magnetic susceptibility for the three boreholes is plotted in Figure 12, which also shows the lithostratigraphical correlation taken from the descriptive logs. The relationship between susceptibility and lithology is clearly seen. The higher susceptibilities are associated with deposits containing volcanic ash layers, and therefore rich in smectite, such as the basal part of

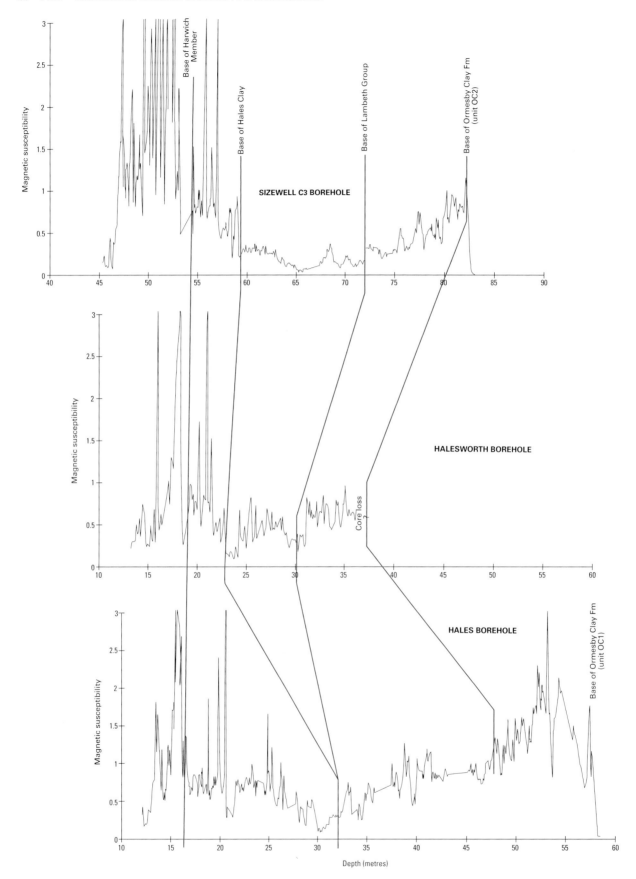

Figure 12   Magnetic susceptibility (nano Tesla) of cored material from boreholes.

the Ormesby Clay Formation (OC1), and the Harwich
Formation (in the latter case the plots are truncated at
3×10^{-3} units for ease of depiction). The apparent susceptibility also reflects the variation in glauconite content
(see Clay mineralogy, below). In the Lambeth Group of
the Sizewell C3 Borehole, there is a clear contrast
between the lower unit, which is mainly clay, and the
upper sandy unit (above about 67 m depth). This pattern
differs from the Halesworth Borehole in which clayey
deposits represent the whole Lambeth Group. The top
metre of the Ormesby Clay Formation in the Halesworth
Borehole has a comparatively reduced magnetic susceptibility, which may possibly indicate oxidation and therefore, perhaps, pre-Lambeth Group weathering.

CLAY MINERALOGY

A full account of the clay mineralogy of these deposits
(except those of the Lambeth Group) is given by
Arthurton et al. (1994). In summary, the clay mineral
assemblage commonly comprises smectite in greater
abundance than illite, with kaolinite and sometimes a
trace of chlorite. However, considerable variation in the
relative proportions of smectite to illite-plus-kaolinite
occur through the sequence.

The lower part of the Ormesby Clay Formation (unit
OC1) is rich in smectite with minor illite and traces of
kaolinite. Glauconite is common in the basal few metres.
In the reddish brown mudstone (lower part of unit
OC2), illite, kaolinite and chlorite increase at the
expense of smectite. Proportions of illite and kaolinite
decline slightly in the upper part of the Ormesby Clay
(OC3 and OC4).

In the Hales Clay Member the trend is reversed.
Kaolinite, illite and traces of chlorite generally increase
upwards. The 'Harwich Member' is characterised by a
high smectite content throughout, but kaolinite and illite
are more abundant than in the Ormesby Clay. Smectite
content is closely associated with the ash layers; where
these are less abundant, proportions of the other clay
minerals rise.

THREE

Late Neogene and Early Quaternary: The Crags and related deposits

During the late Neogene the district became submerged as a part of the marine North Sea Basin, around the margins of which there developed a system of prograding deltas (Cameron et al., 1992). During the Pliocene and Early Quaternary the district lay near to the western coast of this basin, and marine strata of the Coralline Crag and of the later Crag Group were deposited. Opinions vary as to whether the sea receded steadily from west to east during Crag Group times (Arthurton et al., 1994), or whether the sea transgressed and regressed several times (Funnell, 1991): evidence from the present district supports the latter view. Either way, by the end of Crag deposition only the north-eastern corner of the district lay under the sea.

The Cromer Forest-bed Formation, which succeeds the Crag Group, is restricted to the north-east of the district. It comprises marine and estuarine sediments laid down from late Pastonian into Cromerian times. As the sea regressed eastwards, fluvial systems developed upon the landward part of the district. Deposits of two such fluvial systems are preserved: in the south, the Kesgrave Group, comprising river terrace deposits of the proto-Thames, and in the north the Bytham Sands and Gravels. The latter interdigitate with the Cromer Forest-bed in the north-east of the area. Because the sea regressed over a period of time the boundary between marine and fluviatile sedimentation must be diachronous, so that the older, higher terraces of the Kesgrave and Bytham groups may be the same age as the youngest Crag. However, no positive evidence for such diachronism can be cited within the district, largely because of the limited chronostratigraphic evidence available.

It has been traditional Geological Survey practice to separate 'Solid' (bedrock) and 'Drift' (superficial deposits) on maps and in memoirs, the origins of the term 'drift' stemming from the formative years of geological science. The base of the drift is generally taken at the base of the oldest non-lithified deposits of terrestrial origin. In most parts of Britain this boundary is an obvious one, where Quaternary drift overlies much older rocks, but in East Anglia the boundary lies within the Quaternary sequence; the Coralline Crag and Crag Group are regarded as solid, whilst the Kesgrave Group, the Bytham

Sands and Gravels and the Cromer Forest-bed Formation are classed, with all younger deposits, as drift.

CORALLINE CRAG FORMATION

Distribution and lithostratigraphy

The Coralline Crag is a formation of carbonate-rich skeletal sands with an onshore outcrop restricted to south-east Suffolk. Within the district the formation crops out around Aldeburgh, northwards to the headland of Thorpeness, and thence offshore for some 5 km where it has been sampled in vibrocores (Figure 13).

Onshore, the formation rests unconformably on an undulose surface of the Thames Group, reaching a maximum thickness of approximately 25 m, although in the present district the maximum recorded thickness is 15.5 m, proved in boreholes to the north of Aldeburgh. The formation has been subdivided on lithology and biofacies into a sequence of members (Balson et al., 1993). The uppermost of these, the Aldeburgh Member,

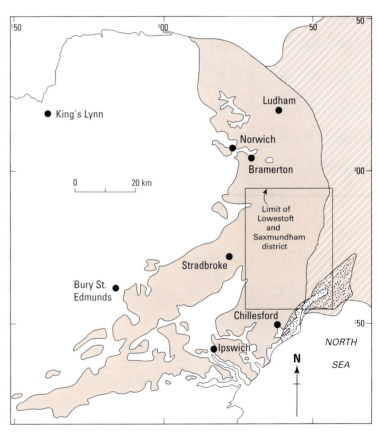

Figure 13 Regional distribution of the Crag Group and Coralline Crag, modified from BGS (1985) and Zalasiewicz et al. (1988).

is exposed at the surface in pits around the town of Aldeburgh, the type area, where in boreholes it has been proved to rest unconformably on the Ramsholt Member. The Ramsholt Member rests unconformably on the Thames Group and contains a basal lag gravel rich in pebbles of phosphatic mudstone largely derived from that formation (Balson, 1989). The Sudbourne Member rests on the Ramsholt Member to the south of the River Alde, but appears to be absent from the present district.

The Aldeburgh Member is up to about 14.5 m thick and consists of a carbonate-rich sand with an average of 65 per cent CaCO$_3$ although samples from surface exposures may contain over 80 per cent CaCO$_3$. Aragonitic shell debris has largely been lost by selective dissolution, so the faunal remains are dominated by bryozoans and calcitic molluscs. Offshore vibrocores have recovered fossiliferous carbonate sands which exhibit the same aragonite leaching as onshore exposures, indicating that the dissolution is not a recent subaerial phenomenon. The acid-insoluble sand fraction is moderately to poorly sorted medium-grained sand; mud content is low, 6.6 per cent on average.

In surface exposures faint horizontal or gently dipping bedding planes can be seen, and in some places it is clear that these are the bounding surfaces of cross-bedded sets up to 30 cm thick. The cross-bedding foresets dip to the south-south-west, a direction similar to that of cross-bedding throughout the formation, implying a constancy of the dominant direction of sand transport throughout Coralline Crag deposition.

The offshore outcrop was noted by Dalton and Whitaker (1886) and Reid (1890) but exposures in the vicinity of Thorpeness were noted as early as 1839 by Sir Charles Lyell. Lyell recorded Coralline Crag at Sizewell Gap but this record has not been confirmed by recent work. He also recorded exposures of Norwich Crag overlying Coralline Crag at low water at Thorpe (now the village of Thorpeness, not the headland). A further record of Coralline Crag at low water to the north of Thorpeness appears to be erroneous, and due to the misidentification of lithified sands of the Crag Group, currently exposed there at Spring tide low water.

Offshore, seismic records show the formation to reach about 25 m in thickness and to extend some 14 km to the north-east of Thorpeness as a partially concealed elongate body. The north-western flank of the outcrop is bounded by a north-west-trending valley which extends landwards beneath Sizewell. The seismic records show a strongly eroded upper surface to the formation and internal reflectors dipping at a low angle to the south.

Large blocks of lithified Coralline Crag are often found on the beach between Thorpeness and Sizewell, and are presumably derived from the subtidal outcrop by marine erosion.

Age, fauna and sedimentation

Estimates of the age of the formation have been based mainly on analysis of macro- and microfossil assemblages in the Ramsholt Member. In particular, studies on nannofossils, planktonic foraminifera, ostracods and

pollen point to a late Early Pliocene age for the member (Balson, 1990). A brief hiatus is indicated by the presence of phosphatic pebbles at the base of the Aldeburgh Member in boreholes and it is likely that this unit is of early Late Pliocene age.

The Coralline Crag contains a rich and abundant fauna despite the dissolution of aragonitic fossils. Bryozoans, which are dominantly calcitic, are conspicuous. Globose colonies of the large cyclostomes *Meandropora aurantium*, *M. tubipora* and *Blumenbachium globosum* (syn. *Alveolaria semiovata*) up to 10 cm in diameter are noted for their abundance in the Aldeburgh Member. These bryozoans lived on the sea floor or occasionally on erect stems of sea grasses or hydroids (Balson and Taylor, 1982). The surfaces of the colonies show relatively little abrasion or colonisation by epifauna. Other bryozoans include globular colonies of *Turbicellepora*, large eschariform colonies of 'Eschara' *pertusa* and *Biflustra savartii*, and well- preserved colonies of *Cellaria* sp. *Cellaria* has a colony consisting of a series of cylindrical calcitic internodes which branch dichotomously at flexible organic nodes. The preservation of the original configuration of the internodes in some colonies is remarkable, considering the coarse bioclastic nature of the enclosing sediment. Bryozoans are also conspicuous members of the encrusting epifauna, together with barnacles and occasional serpulids which are found encrusting calcitic shells or on the surface of moulds of aragonitic shells. The presence of an abundant, well-preserved encrusting epifauna is evidence of relatively slow rates of sedimentation and is characteristic of the Aldeburgh Member.

The molluscan fauna includes shells of calcitic bivalves such as *Chlamys*, *Anomia* and *Ostrea*, the latter often showing clionid borings as further evidence of slow sedimentation and in-situ bioerosion. Aragonitic molluscs such as *Scaphella lamberti*, *Arctica islandica* and *Glycymeris glycymeris* are present as moulds, *G. glycymeris* sometimes occurring as moulds of articulated specimens.

Many of the larger fossils in the Aldeburgh Member are well preserved, indicating that much of the fauna lived in the vicinity. However, the sand which formed the substrate for these organisms consists of extremely abraded and comminuted skeletal debris, indicating transportation from elsewhere. Many of the molluscan shells were encrusted by bryozoans or barnacles or were bored by marine organisms including *Cliona*. This evidence suggests relatively slow rates of deposition. Conversely, the large bryozoan colonies of *Meandopora* and *Blumenbachium* are generally not encrusted or bored, suggesting rapid burial.

The explanation of these apparent contradictions may be that the low-relief bedforms which deposited the low-angle dipping beds were only periodically mobile, perhaps during severe storms. These periodic movements rapidly buried the indigenous fauna with sand, allowing preservation of relatively delicate bryozoan colonies like the articulated *Cellaria*. The lack of epifauna or borings on the surfaces of the large bryozoan colonies may be due to the same process. During intervening periods conditions were more tranquil, with occasional small bedforms migrating across the low-angle bedform towards the south-west, the regional sand transport direction. Low-energy conditions also allowed the accumulation of a thin layer of juvenile pectinid valves, possibly as a result of a mass mortality event.

The Coralline Crag has been proved in the North Warren and South Warren boreholes between Thorpeness and Aldeburgh (see Appendix 2).

CRAG GROUP

All Late Pliocene and Early Pleistocene marine deposits within the district, postdating the Coralline Crag and predating the marine and estuarine sediments of the Cromer Forest-bed Formation, are considered here as the Crag Group. These sediments accumulated towards the western margin of the subsiding North Sea Basin, which is believed to have had similar geometry and tidal characteristics to those of the present day (Anderton et al., 1979), although the sea area must have been somewhat more extensive. The sediments are mainly sands, but include beds of gravel and of clay, particularly towards the top of the group.

Nomenclature

There is a vast literature dealing with the Crag, dating particularly from the 19th century when the high quality of preservation of much of its fauna, its ease of excavation, and its proximity to London, led to widespread scientific study. Reid (1890) gave an exhaustive review of Crag literature, and recently Arthurton et al. (1994) reviewed the more seminal works.

'Crag' was originally an East Anglian dialect term for any sand rich in marine shells; Taylor (1824) appears to have been the first scientist to apply the name Crag in a strictly geological sense although the term had been used earlier by William Smith. Charlesworth (1835) recognised two divisions of the Crag, a Lower or Coralline Crag and an Upper or Red Crag, and later (1837) established a third, higher, division, the Mammaliferous Crag of Norfolk. This latter was referred to as Norwich or Fluvio-marine Crag by Lyell (1839) and the name Norwich Crag has been used for it ever since (Woodward, 1881). Harmer (1898, 1900a, b) proposed the name Icenian Crag to replace Norwich Crag, and although synonymous both terms continued to be widely used.

Prestwich (1849) introduced the terms Chillesford Sand and Chillesford Clay to describe deposits broadly similar to the Norwich Crag, and overlying the Red Crag, in Suffolk. Later workers applied the term Chillesford Clay to virtually any micaceous clay in the Crag of East Anglia, but ultimately the term came to be restricted to one individual clay body centred on Chillesford in the Woodbridge district of Suffolk (Funnell, 1961; Dixon, 1972). The Chillesford Sand has also been termed Creeting Beds (Dixon, 1978) or Creeting Sands (Allen, 1984). Prestwich also (1871b) introduced the term Westleton Beds, to describe a group of lenticular flint gravels within the Norwich Crag, although at the time he

believed them to overlie the Crag rather than to form a part of it.

It is worth emphasising here that the Coralline Crag is not included in the Crag Group of the present terminology. This has come about because the Coralline Crag is a distinctive deposit, always recognisable, whereas the Red and Norwich crags are not always distinguishable, and thus require a single name, Crag Group, in areas of poor exposure (see below).

Stratigraphical framework

A lithostratigraphical classification of what is now called the Crag Group was proposed by Funnell and West (1977) (Table 4), although at that time there was no known means of mapping the various units. Bristow (1983) considered that it was not possible to map boundaries within the Crag, since he could not distinguish the Red and Norwich Crag lithologies by normal mapping techniques in the field. He also pointed out that the Red Crag seen at surface in southern Suffolk is lithologically different from strata of the same age proved in the Ludham Borehole in Norfolk. He suggested that the Red and Norwich crags should be grouped as the Crag Formation, while retaining their chronostratigraphical stage names, biozones etc.

Contrary to the experience of Bristow (1983), later surveyors (Zalasiewicz and Mathers, 1985; Zalasiewicz et al., 1988, 1991), mapping in south-east Suffolk (including a part of the present district) have shown that it is possible to map a boundary between Red and Norwich crags. They defined a Red Crag Formation, with its type locality as the cored sequence between depths of 6.82 m and 26.00 m in the Wantisden Hall Borehole [3601 5215], and subdivided it into the Sizewell and Thorpeness members on the basis of further boreholes.

Zalasiewicz and Mathers (1985) formalised the Norwich Crag Formation with the Chillesford Sand and Chillesford Clay as members. For a type section of the Chillesford Sand Member they chose the cored sequence from 1.42 m to 6.82 m in the Wantisden Hall Borehole. They intended this member to correspond to the Norwich Crag of previous literature. For a type section of the Chillesford Clay Member they chose Chillesford Brickyard Pit [3880 5258], to the south of the present district.

In the present memoir, the terms Red Crag Formation and Norwich Crag Formation are used in the sense defined by Mathers and Zalasiewicz (see above). The base of the Norwich Crag is taken at an unconformity at the base of the Bramertonian/Antian strata (Table 4). In the south of the district, the Red and Norwich Crags are lithologically distinct, and the Red Crag is distinguished on the Saxmundham 1:50 000 map. However, over much of the district these formations could not be separated in the field, since sands and clays stratigraphically below the unconformity, and so belonging to the Red Crag Formation, are lithologically indistinguishable from the Norwich Crag above it. Hence, apart from the Red Crag in the south of the district, all Crag sands are shown on the 1:50 000 maps as undivided Crag Group, as in the Great Yarmouth district (Arthurton et al., 1994).

Table 4 An outline of the nomenclature of the Crag Group and associated strata.

British stages after West (1961, 1980); Funnell (1961); Funnell and West (1977); Beck et al. (1972)	Lithostratigraphical units of Funnell and West (1977)		Revised stages of Gibbard and Zalasiewicz (1988); Gibbard et al. (1991)	Lithostratigraphical units of Zalasiewicz and Gibbard (1988)		Lithostratigraphical units used in this memoir
Cromerian	Bacton Member / Mundesley Member / West Runton Member	Cromer Forest Bed Formation	Cromerian	Bacton Member / Mundesley Member / West Runton Member	Cromer Forest-Bed Formation / Kesgrave Formation	Kesgrave Group
Beestonian	Runton Member		Beestonian	Runton Member		Bytham Sands and Gravels
Pastonian	Paston Member		Pastonian	Paston Member		Cromer Forest-Bed Formation
Pre-Pastonian b–d	Sheringham Member			Sheringham Member		
Pre-Pastonian a	Sidestrand Member		Pre-Pastonian/Baventian	Sidestrand Member	Norwich Crag Formation	Chillesford Clay Member
Bramertonian	Chillesford Beds / Norwich Member (inc. Easton Bavents Clay)	Norwich Crag Formation	Bramertonian/Antian	Westleton Beds / Easton Bavents Clay / Chillesford Clay Member / Chillesford Sand Member		Norwich Crag Formation
Baventian						
Antian						
Thurnian	Ludham Member	Red Crag Formation	Thurnian	Thorpeness Member	Red Crag Formation	Red Crag Formation — Thorpeness Member
Ludhamian	Red Crag Member		Ludhamian	Sizewell Member		Sizewell Member
Pre-Ludhamian			Pre-Ludhamian			unconformity

(Crag Group; Kesgrave Group)

Clay beds within the Crag Group are shown on the 1:50 000 maps where they are sufficiently thick, and the Chillesford Clay Member of Zalasiewicz and Mathers (1985) is distinguished by name on the Saxmundham map. Gravel beds within the Norwich Crag are also shown where possible; these are the Westleton Beds of earlier authors, but that term is used here in a facies sense only. The gravels are scattered through the Norwich Crag, so cannot be regarded as a separate formation, and individual gravel beds are too numerous to be usefully named as members. The relationships between the named components of the Crag Group are shown diagramatically in Figure 14.

Distribution

The present extent of the Crag Group in East Anglia is shown in Figure 13; the original western extent of deposition is uncertain. Nickless (1971) recorded a general westward coarsening of sands in the area south-east of Norwich, and if the sands in question were correctly identified as Crag, this would imply that the coastline lay not far to the west of the present outcrop.

In East Anglia the Crag does not exceed 70 m in thickness (Arthurton et al., 1994), but beneath the North Sea and in the Netherlands beds of the same age are much thicker, reaching 350 m in the Southern North Sea Basin and 800 m in the Central North Sea Basin (Cameron et al., 1992). In the present district the thickness ranges from zero in areas where it has been totally eroded away, to around 70 m; 69.5 m were logged in the Stradbroke Borehole [232 738] just to the west of the district. It is not possible to show the variations in thickness across the district on a simple diagram, partly because of the problem of distinguishing Crag from later arenaceous strata in boreholes, and partly because of the variations in thickness where the Crag has been eroded to form post-Anglian valleys. However, Figure 15 shows contours on the base of the Crag.

The group is present beneath a large part of the district (Figures 13, 15), being absent only where Coralline Crag, Thames Group and Upper Chalk occur at rockhead. However, the western limit of the Crag Group is not well defined since it is based on borehole logs in which it is difficult to distinguish the Crag from overlying sandy deposits. The Crag Group rests uncon-formably upon Palaeogene strata in the east (Figure 15), but oversteps westwards on to Upper Chalk. Following the final eastward retreat of the Early Pleistocene sea, the Crag became wholly buried beneath later Pleistocene fluvial and glacigenic strata up to Anglian in age, but from the late Anglian onwards, erosion exposed the Crag in river valley sides and towards the coast, and it still crops out widely in these situations. It is obscured by later Pleistocene deposits in the valley bottoms, particularly in that of the River Waveney.

Regionally the base of the Crag has been tilted up to the west and rises at about 1 m per km (Mathers and Zalasiewicz, 1988). Figure 15 shows, in addition to this tilt, two north-easterly trending depressions on the base of the Crag, one passing through Bungay and the other through Leiston and Southwold. These lead to considerable variations in the thickness of the group.

In the south of the district, northwards from Aldeburgh, Zalasiewicz et al. (1988) demonstrated a slope on the base of the Crag Group of 3 to 5°, between boreholes in which thin Norwich Crag rested on Coralline Crag and then on Thames Group at 15 m below OD, and boreholes in which a thick sequence of Norwich and Red Crags rested directly on Thames Group at 40 to 50 m below OD. The lack of any evidence for relative movement within the Palaeogene or Cretaceous strata underlying the Crag, and the absence of Coralline Crag in the deeper part of this structure, led them to conclude that the slope is erosional, representing an uneven topography at the time of Crag deposition, the strata having accumulated more thickly in the depression than elsewhere.

The thick sequence in the north-west of the district lies within the Stradbroke Trough (Funnell, 1972), which passes north-eastwards beneath Bungay (Figure 15). Bristow (1983) noted that this basin has a sharp south-eastern boundary, and that contours on the base of the Crag within the basin show elongate depressions trending north-north-east. He used these features as evidence that Crag sedimentation within the basin was controlled by contemporaneous faulting. Of the faults which he named, the eastern ends of the Bedingfield, Debenham and Pettaugh faults appear on Figure 15. Contemporaneous subsidence of the sea floor would explain the presence of some 69.5 m of Crag, all of shallow-water facies, in the Stradbroke Borehole [232 738]. Bristow's interpretation is supported by a gravity survey (Cornwell, 1985) which indicated basement faults underlying two of his postulated faults. Similar structural features are known elsewhere; fault-bounded basins infilled with Quaternary sediments occur in the Netherlands (Van Staaldvinen et al., 1979), and near-surface expressions of deep-seated faults have been described in the Quaternary of the southern North Sea (Balson and Cameron, 1985).

Earlier it had been suggested that the depressions in the sub-Crag surface may have been formed by fluvial erosion (Carr, 1967). Alternatively, Funnell (1972) likened the topography of the basal Crag surface within the Trough to the present sea floor of the English Channel and favoured tidal scour as the mechanism of

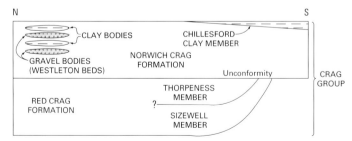

Figure 14 Diagram to show relationships between the lithostratigraphical components of the Crag Group. The unconformity relates to the Bramertonian transgression.

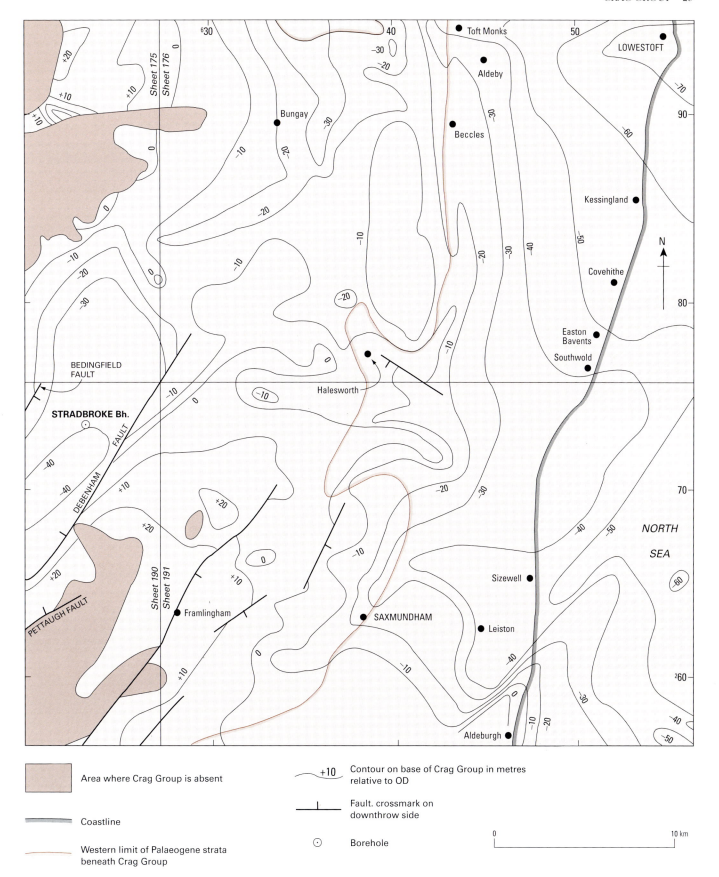

Figure 15   Contours on the base of the Crag Group. The diagram extends to the west of the district to include the Stradbroke Borehole.

formation, as did Zalasiewicz and Gibbard (1988). In practice it is possible that more than one process has been involved. However, Figure 15 shows the Stradbroke Trough and the trough beneath Leiston to be not only parallel but also very similar in shape, with steep south-east flanks and shelving north-west flanks. The implicaton is that both troughs may be structurally controlled, with faults on their south-east sides.

Figure 2 shows the distribution at rockhead of the Red Crag where it can be mapped, and of the mappable clay and gravel beds within the Norwich Crag. The Norwich Crag is present throughout the district, but the Red Crag is known only in areas where the Crag Group is thick, notably in the Stradbroke Trough and in the area from Aldeburgh to Southwold. Thus it is possible that the Red Crag was only deposited in structurally controlled troughs, while the Norwich Crag rests unconformably on Red Crag in the troughs but on Palaeogene strata or Upper Chalk over the intervening highs. Bristow (1983) suggested that the faulting which bounded the Stradbroke Trough had ceased before the Bramertonian transgression.

The Chillesford Clay was formalised as a member of the Norwich Crag Formation by Zalasiewicz and Mathers (1985) and Zalasiewicz et al. (1991). They chose a type section at Chillesford Brickyard [3880 5258] in the Woodbridge district (sheet 208), from a list of sites described by Prestwich (1849). It is an elliptical lenticular body, 12 km by 4 km, trending north-east and dipping gently in the same direction. It extends into the present district as far as a point [461 583] north of Aldeburgh where it rests on Norwich Crag sands at 10 m above OD.

It is up to 5 m thick, has a transitional contact with the underlying sands, and forms the top of the Crag in the south of the present district.

Clays lithologically similar to the Chillesford Clay and of the same Baventian age are found along the coast from Easton Bavents to north of Covehithe [515 784 to 534 836] (Plates 3, 4). Several closely spaced lenticular bodies are recorded, overlain by sands with gravel beds. At Thorington [423 728] a similar clay body occurs sandwiched between two gravels. Earlier (Pre-Ludhamian and Ludhamian) clay bodies up to 9.9 m thick were recorded at depth in the Stradbroke Borehole [232 738] just to the west of the district.

A bed of gravel or gravelly sand commonly occurs at the base of the Crag, but is more strongly developed at the base of the Red Crag in the south-east than at the base of the Norwich Crag in areas where the Red Crag is absent. Lenses of flint gravel higher in the Norwich Crag, traditionally termed Westleton Beds, are strongly concentrated in the area between Southwold [50 76], Halesworth [38 78], and Dunwich Heath [47 67] (Figure 2). Particularly fine sections can be seen in the sea cliffs at the last-named locality. At Thorington [423 728] and at Covehithe, these gravels are found interbedded with clay seams. Thinner developments of similar gravels, significantly containing low concentrations of vein quartz and quartzite, are found in other areas: for instance, beds up to 2 m thick were recorded in boreholes in the upper Waveney valley (Auton et al., 1985). These gravels may be restricted to the upper part of the Crag (i.e. not below the Baventian or possibly the Antian–Bramertonian), but palaeontological confirmation is lacking.

Plate 3   Clay within the Norwich Crag Formation exposed as a bench on the foreshore at Covehithe; overlain by Norwich Crag sands and Corton Formation (GS 562).

Plate 4 Detail of clay within the Norwich Crag Formation at Covehithe, showing bioturbation (GS 563).

Stratigraphy and sedimentation

The **Crag Group** dominantly comprises fine- to coarse-grained micaceous sands. Where these are unweathered, below the water-table, they are dark green in colour owing to their high glauconite content. However, where they are seen in a weathered state the sands are yellowish to reddish brown in colour and yield a light sandy soil. Augering or excavation reveals these weathered sands to contain beds of ferruginous concretions (iron pan), which have formed from the iron oxides and hydroxides released by the weathering of glauconite; such concretions are often found ploughed up in the soil.

Where the **Red Crag** rests upon the Upper Chalk there is commonly a basal bed up to 2 m thick of pebbles and cobbles of glauconite-coated flint. This bed represents a transgressive gravel beach deposit and is believed to be reworked from the Bullhead Bed at the base of the Thanet Formation (Bristow, 1983; Arthurton et al., 1994), since the glauconite coating of the flints is so similar in both cases (Merriman, 1983). Where the Crag

overlies Palaeogene strata the basal bed commonly contains a high proportion of phosphate pebbles (Dixon, 1979; Bristow, 1983). Quartz and rare quartzite pebbles also occur. However, this basal gravel is not universally developed and in some cases the basal bed of the Crag Group comprises sand or clayey sand.

Overlying the basal pebble bed, the Red Crag comprises poorly sorted cross-bedded medium- to coarse-grained shelly sands, with a gradual coarsening-upward trend. These are believed to represent an accumulation of material transported in tidal sand waves (Dixon, 1979; Zalasiewicz and Mathers, 1985; Mathers and Zalasiewicz, 1988). The cross-bedding sets may be up to about 5 m in height (Dixon, 1979), most commonly 2 to 3.5 m; they imply unidirectional flow in water depths of up to 25 m.

North of Aldeburgh, Mathers and Zalasiewicz (1988) and Zalasiewicz et al. (1988) distinguished two units within the Red Crag. The lower, the **Sizewell Member**, comprises 13 m of medium- and coarse-grained greyish green shelly sands interbedded with clays containing thin silt and sand laminae. The sands are moderately to poorly sorted and contain much glauconite and broken molluscan shell debris as well as quartz. This member is considered to be subtidal in origin. The overlying **Thorpeness Member** is 20 to 30 m thick and comprises two coarsening-upward cycles of shelly fine- to medium- and rarely coarse-grained sand with a few laminae and thin beds of silty clay. The coarser sands include fragmented molluscan material, rip-up clasts of silty clay, and rare phosphate pebbles. This member is interpreted as indicating the growth or migration of very large-scale bedforms such as tidal sand ridges, and hence represents sedimentation in shallower water than the Sizewell Member.

The northward extent of the Sizewell and Thorpeness members is unknown since little is known of Red Crag sedimentology in the northern parts of the district. However, it is notable that in the Ormesby Borehole, to the north of the present district, sediments belonging to the Red Crag Formation (identified in the borehole as Lithofacies Units 1 and 2 by Arthurton et al., 1994), are similarly interpreted as of subtidal and intertidal origin, with the same shallowing-upward trend, although the sediments themselves are generally more argillaceous than those in the present district.

A different facies of the Red Crag was proved to the west of the district, in the Stradbroke Borehole [2328 7383] (Figure 15) (Lord, 1969). A thickness of 69.5 m of Crag comprised green and grey sands, fine-grained above but becoming coarser and more shelly with depth. Clay beds up to 9.9 m thick were present throughout, as well as 'claystone' nodules. Similarity with the clay beds of the Norwich Crag would imply deposition of the Red Crag here in much shallower water than that inferred for the Sizewell and Thorpeness members.

The **Norwich Crag** comprises a widespread sheet of well-sorted fine- to medium-grained sand, locally including beds of clay (Chillesford and other clays) and gravel ('Westleton Beds'). In south-east Suffolk the sands have been called the Chillesford Sand Member by Mathers and Zalasiewicz (1985). Small-scale sedimentary structures in the sands include horizontal bedding, bipolar ripples, flaser bedding, isolated thin clay drapes, trough cross-sets up to 0.3 m high, vertical burrows, channel scours and polygonal mudcracks. Deposition in a tidal flat environment is inferred (Dixon, 1972; Allen, 1984; Zalasiewicz and Mathers, 1985; Mathers and Zalasiewicz, 1988; Zalasiewicz et al., 1988).

Shelly sands are rare in the Norwich Crag; it has been suggested that their scarcity is due to decalcification, but this is questionable, since trace fossils are also rare, and shells do sometimes appear in the weathered sands found at outcrop. Between Aldeburgh and Sizewell (Zalasiewicz et al., 1988) the Norwich Crag sands are moderately to well sorted and fine to medium grained, with isolated shelly lenses of fragmented molluscan valves, silty clay laminae and clay intraclasts; possibly the presence of clay aids preservation of the shells. Foresets of cross-bedding commonly dip at more than 20°, and cross-set thickness is usually less than 0.3 m.

The marine (beach, estuarine or deeper-water) character of the Norwich Crag has been confirmed by electron-microscope examination of the sand grains (Krinsley and Funnell, 1965), although with an aeolian episode for some of the grains, which have presumably been blown into the marine environment.

The Chillesford Sand at Chillesford Church Pit [3828 5230] has yielded 22 species of ostracod. The assemblage is characterised by *Baffinicythere howei* Hazel, *Cytheropteron nodosalatum* Neale and Howe, *Leptocythere psammophila* Guillaume, *Finmarchinella logani* (Brady and Robertson)

and *Thaerocythere mayburya* Cronin. The ostracod fauna is indicative of fully marine conditions, a frigid environment with mean summer temperatures of about c.3°C and mean winter temperatures of about -1.5°C.

The gravel bodies referred to collectively as **Westleton Beds** (Frontispiece; Plate 5) are dominated by chatter-marked high-sphericity flint pebbles and cobbles. These are normally grey or brown both inside and out, and are believed to be second-cycle material derived from pre-existing Palaeogene deposits. Hey (1967) demonstrated that the 16–32 mm fraction contains 96 to 99 per cent flint, of which 53 to 78 per cent are rounded. Lithologies other than flint are rare in pebbles greater than 32 mm. 'Box-stones', with scattered grains of quartz in a phosphatic groundmass, are recorded at Easton Bavents (Hey, 1976); these may be derived from the Palaeogene deposits or, less likely, from the Red Crag. Amongst the far-travelled components, quartz and colourless or yellow quartzite are the commonest. Hey (1976) recorded a sub-angular block of coarse white quartzite 23 cm long at Easton Bavents. Prestwich (1871b, 1890) suggested that these erratics had been transported from the Ardennes by some marine process, but a more likely source is the Triassic Kidderminster Formation ('Bunter Pebble Beds') of the Midlands (Hey, 1976), and transport by rivers.

Sinclair (1990) found that 6 to 10 per cent of the flint pebbles, which he termed 'spicular flints', were not derived from the Chalk of Norfolk or from the Palaeogene pebble beds of the London Basin. He considered that these came from the Welton and Burnham

Plate 5
Well-rounded flint gravels (Westleton Beds) within the Norwich Crag Formation at Blyth River Gravel Pit [4119 7672] near Halesworth (GS 564).

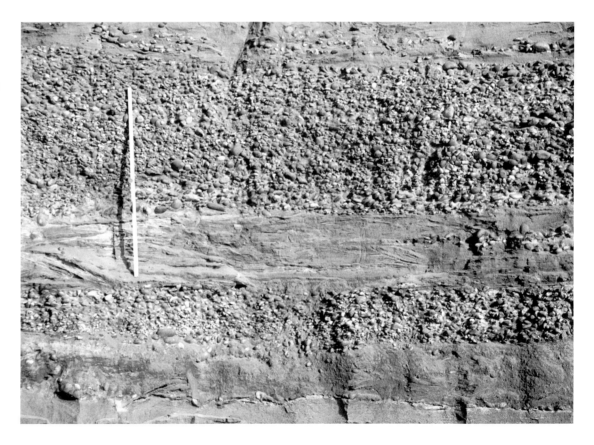

Scale is 1 m

formations of the Lincolnshire Chalk and that these, along with distinctive *Rhaxella* chert from the Corallian of Yorkshire, which consists almost entirely of minute globular sponge spicules, were transported into the Crag by longshore drift. This seems unlikely, however, since any intervening estuaries would interrupt the flow of the gravel. Hey (1976) suggested that the *Rhaxella* chert was transported from Yorkshire by ice, since Funnell and West (1962) consider that ice sheets were present not far from East Anglia during the cold phase (Baventian) during which the Easton Bavents deposits were formed, but this seems equally unlikely since a more varied suite of erratics might then be expected. Fluvial transport would appear to be the most likely explanation, as in the case of the quartz and quartzite.

The gravels are well sorted: Hey (1967) found in a sample from Westleton that over 70 per cent comprised pebbles of 10 to 50 mm mean diameter. They also show a high degree of roundness and sphericity. Fossils are rare and are generally very worn, but include elephant bones and whale vertebrae. Hey (1967) recorded impressions of marine molluscs, confirming the marine origin of the deposits. He noted that the gravels in the Southwold–Halesworth–Dunwich area occur as large-scale planar cross-stratified units, in sets up to 10 m thick with foresets dipping south-eastwards at up to 10°. By comparison with the present-day Dungeness coastline he interpreted the structures as beachface deposits of a coastline that was prograding towards the south-east.

Later work by Sinclair (1994) and Mathers and Zalasiewicz (1996) confirmed the beachface origin of the main gravel bodies, but also recognised further facies, one characterised by horizontally stratified sands and steeply incised gravel-filled channels, and another by regularly spaced channels up to 2 m deep and filled with cycles of gravel fining upwards into sand. They interpreted the three facies as respectively gravel-dominated beachface deposits, sandy shoreface deposits, and an offshore facies dominated by storm-generated sedimentation in rip-current channels. Mathers and Zalasiewicz (1996) interpreted these facies as laterally equivalent and eastward-deepening, but at the Quay Lane Pit, Reydon [483 775] (Figure 16) they indicate that the three facies occur superimposed on one another, which must indicate a deepening-upward relationship.

The **Chillesford Clay**, up to 5 m thick, is typically an unfossiliferous pale to medium grey or buff silty clay, with ochreous staining, scattered poorly defined laminae of silt and sand, and locally bands of fine sand with clay laminae. The clay seam which crops out on the foreshore at Easton Bavents and Covehithe, and which is less than 3 m thick, is lithologically similar. It contains laminae and some thin beds of fine-grained sand, locally with ripple cross-lamination. West et al. (1980) recorded evidence of intermittent desiccation at Covehithe, and palaeontological evidence of subarctic conditions.

The lithology of the clay bodies, including the evidence of desiccation, indicates a high intertidal or estuarine environment. Furthermore, the situation of the clay bodies, immediately below gravels at Covehithe and between two gravel beds at Thorington (Plate 6), implies a close genetic relationship between the two lithologies: it is suggested that the clays formed in an estuarine environment protected on the seaward side by banks of beachface gravel. The Chillesford and Easton Bavents–Covehithe clays are of approximately the same age but are not apparently connected to one another, suggesting that they represent the deposits of two separate estuaries. All the major constituents of the gravels, but not the 'spicular flints' or the *Rhaxella* chert, could be derived from pre-existing strata to the west of the present district, and fluvial transport from that direction would seem to be the most likely method of introducing the material into the area.

The geographical concentration of gravels in the Sizewell–Halesworth–Dunwich area and the relationship of the gravels with estuarine clays at Thorington and Covehithe suggest that the river transporting the gravels entered the sea from the west in that area. Hey (1976) considered that a 'proto-Trent' river was the main conduit for the gravel, but the absence of significant gravel beds farther north suggests that little material came via that route, and the absence of gravels associated with the Chillesford Clay suggests that little came via a 'proto-Thames'. The 'spicular flints' and *Rhaxella* chert may have been transported by the same river as the flints, having been disgorged into that river by a north-bank tributary.

The suggestion of a river entering the area from the west is supported by the palynology of the clays (Riding, 1995a, 1995b, 1995c). Samples from sand directly below the clay at Easton Bavents yielded reworked Carboniferous and Jurassic–Cretaceous spores, while reworked elements in samples from the clay at the same site [5147 7840] were dominated by Carboniferous (Dinantian–Namurian) spores and Jurassic (Pliensbachian to Upper Jurassic) miospores and dinoflagellate cysts, with lesser levels of Cretaceous and Palaeogene reworking and very little from the Neogene. Two samples from a clay bed lying between Westleton Beds gravels at Thorington Pit yielded a similar microflora with Carboniferous, Jurassic, Lower Cretaceous and Palaeogene palynomorphs but no Neogene forms. These derived assemblages could only come from the west or north-west; the rarity of Neogene forms rules out derivation from the North Sea area.

Solomon (*in* Funnell and West, 1962, p.135) considered that the heavy minerals of the Crag clays at Easton Bavents and Covehithe were derived from a North Sea Drift ice-sheet, because of their high proportion of alkali amphibole and the common occurrence of sphene. However, Zalasiewicz et al. (1991) disputed this conclusion. They analysed samples from the Chillesford and Easton Bavents clays, from sands underlying the Chillesford Clay, and from the Red Crag, and found sufficient similarity between them to demonstrate a common provenance.

Hallsworth (1994) analysed heavy minerals from nine samples of Norwich Crag sands. The mineral suite was dominated by stable grains (44 to 76 per cent), and the proportion of opaque minerals varied between 24.0 and 58.5 per cent. However, the species range was diverse: 19 different minerals were identified including the unstable species amphibole, sillimanite and apatite. Considerable

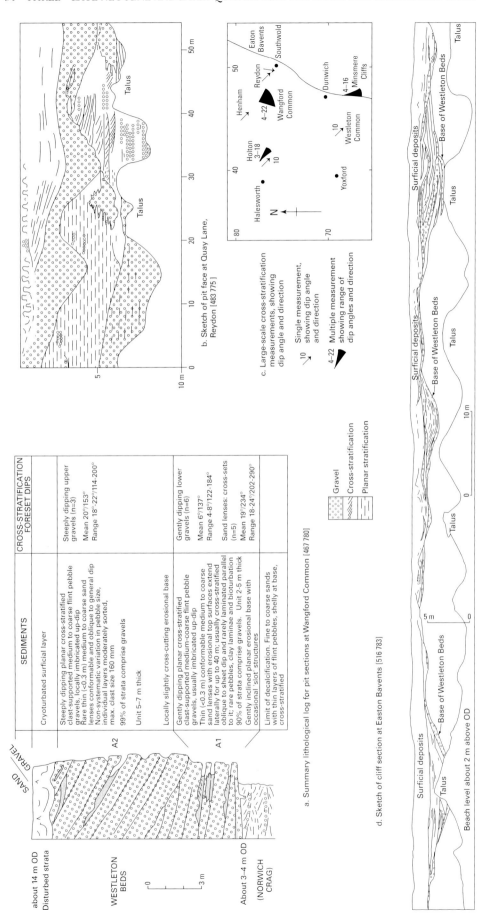

Figure 16 Sketch sections of Norwich Crag gravels (Westleton Beds) and measurement of large-scale cross-stratification (from Mathers and Zalasiewicz, 1996).

Plate 6 Seam of clay (about half way up plate, at top of vertical face) within well rounded flint gravels (Westleton Beds) exposed at Thorington Gravel Pit [423 728] (GS 565). Exposed face is 2 to 3 m high.

variation in the proportion of unstable minerals was probably caused by variations in recent weathering history since the samples were all collected from the weathered zone. Consistent ratios between individual heavy mineral species suggest that the original mineralogy of the Crag was fairly homogenous. All samples but one contained a garnet suite composed of a relatively heterogenous group of generally low-pyrope garnets, characteristic of low to middle grade metasediments, but one sample [from 4238 7281] contained a very diverse assemblage spread over pyrope-poor and pyrope-rich garnets and typical of low to upper amphibolite grade metasediments. Within the Palaeogene deposits of the London Basin these suites are found respectively in the Thanet Formation and the London Clay.

The high proportion of unstable minerals would appear to rule out derivation from the mature Palaeozoic and Mesozoic sedimentary rocks of the Midlands, or from the Lambeth Group of the Palaeogene province of the London Basin. Fresh samples from the Thanet Formation also yield a proportion of unstable minerals rather lower than that of the Crag. However, the sands within the London Clay have a high proportion of unstable minerals, as do the coarser-grained late Palaeogene Bagshot and Claygate Beds. Also, the garnet geochemistry of the Crag is similar to that of the Thanet Formation and London Clay. Thus the Thanet Formation, London Clay, Bagshot and Claygate Beds would appear to be the most likely sources for the Crag. Alternatively, the Crag, Thanet Formation, London Clay, Bagshot and Claygate Beds may derive from a common source of first-cycle material, but this would need to be a metasedimentary terrain such as the Moine or Dalradian of Scotland, and Crag palaeogeography does not support this hypothesis.

Chronostratigraphy and biostratigraphy

Harmer (1898, 1900a, 1900b) established a sequence of molluscan zones and stages, which implied a northward younging of the strata and cooling of the climate. His scheme remained in use until the work of West (1961b) and Funnell (1961). Funnell (1961) examined foraminifera from a water supply borehole at Ludham, 25 km north of the present district, and found that assemblages from the lower part of the Crag resembled those of the Red Crag of Suffolk but not those of the Norwich Crag, while those from higher in the sequence resembled those from surface outcrops of Norwich Crag. The highest faunas resembled those from high in the type sequence at Bramerton Common, and his B II Horizon, which immediately underlies clay at Bramerton, indicated a climate too cold to correlate with the Chillesford Sand of Suffolk.

West (1961b) examined pollen from a Royal Society borehole sunk close to the Ludham water supply borehole. He named four climatic stages, in upward sequence: Ludhamian (warm), Thurnian (cold), Antian (warm) and Baventian (cold) (Table 4). He correlated these with the foraminiferal horizons of Funnell (1961), which similarly indicated alternations of temperature. Further climatic stages were later proposed: Pre-Ludhamian (cold; Beck et al., 1972); Bramertonian (warm; Funnell et al., 1979), Pastonian (warm; West and Wilson, 1966), and the substage Pre-Pastonian a (cold; Funnell et al., 1979; West, 1980) (Table 4). A comprehensive summary of these stages, which are all based on pollen assemblage biozones (West, 1961a, 1980; Beck et al., 1972; Funnell and West, 1977; Funnell et al., 1979) is given by Funnell (1987). However, several workers (notably Bowen et al., 1986; Funnell, 1987; Zalasiewicz and Gibbard, 1988; Gibbard et al., 1991) pointed out that the relationships between some of the stages were unclear and that there might be duplication of stages. In particular it is now suggested (Zalasiewicz et al., 1991) that the Bramertonian predates rather than postdates the Baventian, so that the Antian and Bramertonian represent a single warm stage; and that the Baventian and Pre-Pastonian a represent a single cold stage (Gibbard et al., 1991) (Table 4).

Early workers assigned the Crag to the Pliocene, but on the recommendation of the 28th International Congress (1948) it was placed in the Pleistocene. Mitchell et al. (1973) referred both the Red Crag and Norwich Crag to the Lower Pleistocene, but Funnell (in Curry et al., 1978) considered that a part of the Crag Group was Pliocene. The Plio-Pleistocene boundary was re-positioned at about 1.64 Ma by Aguire and Passini (1985), and Funnell (1987) placed stages up to the Baventian in the Pliocene and younger stages in the Pleistocene. However, the exact position of the Plio-Pleistocene boundary is still in dispute, and Anglo-Dutch practice is to place it at the base of the Pre-Ludhamian, at about 2.3 Ma (Gibbard et al., 1991). This puts all the Crag Group of the present district in the Pleistocene. A further major problem with correlation of the Crag is the presence of large local breaks in the sequence, now believed to be due to an unconformable relationship between the Norwich Crag and the Red Crag.

Pre-Ludhamian and Ludhamian

The Pre-Ludhamian was defined by Beck et al. (1972) on the basis of the pollen sequence in the Stradbroke Borehole [232 738] (Lord, 1969). Here, sediments of this age occur between 15 and 39.4 m below OD, overlain conformably by 45 m of Ludhamian deposits. The Sizewell Member of the Red Crag, which is normally magnetised, was assigned a Pre-Ludhamian age on the basis of pollen, foraminfera and dinoflagellates, while the Thorpeness Member, which is reverse magnetised, is considered to be Ludhamian on the basis of foraminifera (Zalasiewicz et al., 1988).

Thurnian

Deposits of Thurnian age are absent from around Aldeburgh and Stradbroke, but have been identified in boreholes at Sizewell (Funnell 1983b), at Reydon near Southwold (West and Norton, 1974) and between Beccles and Lowestoft (Harland, 1984; Hopson and Bridge, 1987). Thick clays in these sediments at Reydon imply estuarine conditions and may indicate a regression of the sea, in which case Thurnian sediments may never have been deposited as far west as Stradbroke.

Antian–Bramertonian

A major transgresson occurred during the Antian–Bramertonian (Funnell et al., 1979; Bristow, 1983); at Bramerton near Norwich, sediments of this age rest upon Chalk. The Norwich Crag in the south of the district is considered to be Bramertonian (Zalasiewicz et al., 1988); it overlies Thurnian sediments at Sizewell (Funnell, 1983b) but farther south comes to rest upon the Ludhamian Thorpeness Member between Sizewell and Aldeburgh (Zalasiewicz et al., 1988), on the Pre-Ludhamian Sizewell Member between Aldeburgh and Orford, and on Pliocene Coralline Crag just north of Aldeburgh. Norwich Crag sands of Antian–Bramertonian age are also widespread in the north of the district, being recorded at Easton Bavents (Funnell and West, 1962), in boreholes between Beccles and Lowestoft (Hopson, 1991), at Bulcamp [4420 7545] (Funnell, 1983a), and at Aldeby [431 930] (Norton and Beck, 1972). The sands are relatively thin, since the overlying Baventian gravels (Westleton Beds) lie near to the base of the local Norwich Crag; for instance, west of Bramfield only 8.5 m of sands were proved between the Upper Chalk and the lowest gravel of the Westleton Beds.

Pre-Pastonian–Baventian

The type section for the Baventian cold stage lies within the district at Easton Bavents (Funnell and West, 1962; Mitchell et al., 1973). Here Funnell and West (1962) studied pollen and foraminifera, and identified sublittoral shelly sands of Antian age and cool-temperate climate, overlain by clay of Baventian age accumulated in a very cold climate. West et al. (1980) described the pollen spectra, foraminifera and molluscs from similar clays at Covehithe. The pollen indicate cold heathland conditions onshore, with *Empetrum* grass, while the foraminifera and molluscs indicate boreal to arctic conditions; they concluded that these Baventian sediments represent the first cold stage of truly glacial intensity in the English Early Pleistocene sequence. However, Harland (1993) examined Crag sand samples from Covehithe [527 819] for dinoflagellate cysts. Two of these samples were dominated by *Operculodinium israelianum* (Rossignol) Wall, a species now restricted to marine waters warmer than the present-day North Sea (Harland, 1983).

The Easton Bavents clay near Southwold [518 787], contains rare ostracods including the cold water species *Baffinscythere howei*, together with *Cythere lutea* Müller, and *Pontocythere* sp. (Lord et al., 1988).

Zalasiewicz et al. (1991) suggested that the Chillesford Clay is of Baventian to Pre-Pastonian a age, and in part at least equivalent to the clays at Easton Bavents and Covehithe. Pollen identified were largely non-arboreal and indicate a climatic deterioration from cool oceanic to cold, while foraminifera indicate a decline from temperate to cool. However, as at Covehithe, dinoflagellate cyst floras indicate warmer conditions than the other biological indicators. This conflict of evidence has not been satisfactorily explained.

Funnell et al. (1979) associated the Westleton Beds gravels with the Bramertonian transgression, and the estuarine clays of the Baventian cold phase with a regression. However, the interpretation of the major gravel bodies as part of a prograding beachface complex comparable with that at Dungeness (Hey, 1967; Mathers and Zalasiewicz, 1996) rules out any association with a transgression, since the gravels clearly prograded in a seaward direction (south-eastwards), while the Bramertonian transgression would have advanced towards the west. On the contrary, the present survey indicates that the Westleton Beds are interbedded with the estuarine clay bodies, and hence also of Baventian age, and it would appear that between them they formed a coastal complex of shoreface gravel bodies and estuarine mudflats, at a time of relatively constant sea level, lower than that during the Antian/Bramertonian.

Pastonian

At Reydon, the gravel-dominated Westleton Beds beachface deposits are overlain by sandy shoreface deposits and then by an offshore facies dominated by storm-generated sands and gravels in rip current channels (Mathers and Zalasiewicz, 1996). This stratigraphical relationship clearly indicates an upward-deepening sequence. At Thorington and Holton pits the beachface gravels are similarly overlain by sands, while in the Ormesby Borehole, Baventian sands and clays are overlain by a thick sand sequence (Arthurton et al., 1994). If the gravel-dominated shoreface deposits, and the associated clay bodies, formed at a time of stillstand during the Baventian cold stage, as suggested above, this upward deepening trend could represent a marine transgression during the succeeding Pastonian temperate stage. West (1980, p.83) recorded marine sands and clays of Pastonian age at Corton in the Great Yarmouth district, while tidally laminated sediments of Baventian to Pastonian age overlie Baventian silts and clays between Beccles and Lowestoft (Hopson, 1991).

Offshore correlation

The Red Crag Formation of offshore areas (Cameron et al., 1984) may not correlate with its onshore namesake (Cameron et al., 1992). The offshore Red Crag is normally magnetised so could only be equated with the lower part (Sizewell Member) of the onshore Red Crag. Cameron et al. (1984) placed the offshore Red Crag in the Waltonian, which would make it entirely older than the Red Crag of the present district. The offshore Red Crag is overlain unconformably by the Westkapelle Ground Formation, which has reversed magnetisation and Thurnian pollen spectra (Cameron et al., 1984, 1992). This can thus be readily equated with the uppermost (Thurnian) part of the onshore Red Crag, and by extrapolation also with the Thorpeness Member. However, if the above correlations are correct then there should be an unconformity between the Sizewell and Thorpeness members. No such unconformity has been detected, so it may be that the whole of the onshore Red Crag correlates with the Westkapelle Ground Formation (Table 5). The seismic reflector beneath the Westkapelle Ground Formation would thus equate to the unconformity beneath the Red Crag in the present district. This interpretation would, however, conflict with the magnetic evidence.

The Westkapelle Ground Formation is overlain unconformably by the Smith's Knoll and Ijmuiden Ground formations, which derived sediment respectively from Britain and the Netherlands (Cameron et al., 1992). The Smith's Knoll Formation contains dinoflagellate-cyst and pollen assemblages resembling those of the Antian onshore, and hence can be equated with the Antian–Bramertonian Norwich Crag, and by extension probably also that of Baventian age, with the unconformity beneath the Smith's Knoll Formation equating to that beneath the Norwich Crag. A Baventian–Pre-Pastonian mollusc fauna has been recovered from the Crane Formation, an offshore sand bar formed contemporaneously with the deltaic sediments of the Westkapelle Ground and Smith's Knoll formations (Cameron et al., 1994).

The Smith's Knoll and Crane formations are overlain unconformably by the Winterton Shoal Formation, a deltaic unit that contains sediment of both British and Dutch derivation. This may be correlated with the Pastonian marine sediments onshore. Further offshore, deltaic units (Markham's Hole, Outer Silver Pit, Aurora and Yarmouth Roads formations) indicate shorelines east of the present district and hence cannot be correlated with the onshore Crag Group, but must correlate with the Cromer Forest-bed Formation, Kesgrave Group and Bytham Sands and Gravels.

Details

Owing to the difficulty of distinguishing Norwich Crag from Red Crag in boreholes and outcrops in the northern and western parts of the district, the two formations are here considered together.

Waveney valley upstream from Bungay

During the course of an industrial minerals survey of the Waveney valley upstream of Bungay (Auton et al., 1985), 86 boreholes penetrated the Crag Group, of which 42 were sunk in the present district. They proved fine- to medium-grained sands, locally clayey or silty, and greyish, yellowish, and olive-green in colour; glauconite and phosphate grains, bivalve fragments, iron pan and flint pebbles were present. Layers of silt and clay were dark greenish grey in colour and contained glauconite, mica and shell fragments.

Table 5 Correlation of lithostratigraphy between the present district and the Southern North Sea.

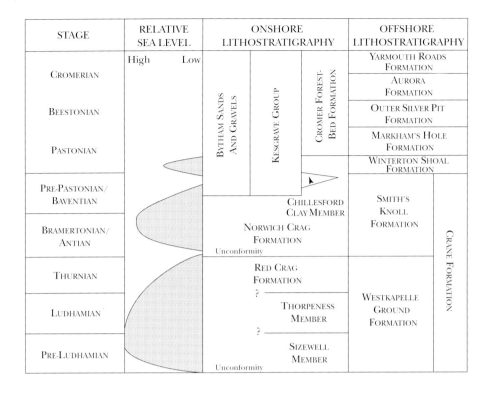

STAGE	RELATIVE SEA LEVEL (High — Low)	ONSHORE LITHOSTRATIGRAPHY	OFFSHORE LITHOSTRATIGRAPHY
CROMERIAN		Bytham Sands and Gravels / Kesgrave Group / Cromer Forest-Bed Formation	YARMOUTH ROADS FORMATION
BEESTONIAN			AURORA FORMATION
			OUTER SILVER PIT FORMATION
PASTONIAN			MARKHAM'S HOLE FORMATION
			WINTERTON SHOAL FORMATION
PRE-PASTONIAN/ BAVENTIAN		CHILLESFORD CLAY MEMBER	SMITH'S KNOLL FORMATION
BRAMERTONIAN/ ANTIAN		NORWICH CRAG FORMATION — Unconformity	
THURNIAN		RED CRAG FORMATION	
		?	WESTKAPELLE GROUND FORMATION
LUDHAMIAN		THORPENESS MEMBER	
		?	
PRE-LUDHAMIAN		SIZEWELL MEMBER — Unconformity	CRANE FORMATION

Particle size analyses of samples lacking glauconite or shell debris gave an average result of 10 per cent fines, 87 per cent sand and 3 per cent gravel. In other samples, pebble counts demonstrated that up to 84 per cent of the gravel fraction was shell debris and iron pan. Flint gravels up to 2 m thick, with up to 60 per cent by weight of the clasts coarser than 4 cm, were recorded in some of the boreholes, mostly in the Diss district to the west. These largely comprised well-rounded black and grey flints, with some subangular flint, and with vein quartz and quartzite varying from zero up to respectively 10 and 17 per cent of the gravel fraction.

Five boreholes penetrated the base of the Crag, which rested upon Chalk. The basal bed of the Crag varied between greyish olive-green shelly glauconitic sand (borehole TM 39 SW/46 [3248 9065]) to coarse gravel with well-rounded glauconite-coated black flints, subangular black flints, shells, iron pan, vein quartz and quartzite (borehole TM 28 SW/48 [2289 8415] just west of the district). West of Bungay the Crag reaches maximum thicknesses of only around 5 m on sheet TM 29 SE and 15 m on sheet TM 39 SW, thickening into the Stradbroke Trough to at least 25 m on TM 39 SE east of Bungay and TM 28 SE to the south-west, 27 m on TM 38 NW around Bungay and 41 m on TM 38 NE to the south-east. There are no natural outcrops in this area, but dark green glauconitic slightly clayey fine-grained sand containing a high proportion of white shell fragments has been dredged from below drift at Flixton Quarry [294 862].

Beccles to Lowestoft

Boreholes drilled in 1983 as part of a BGS regional study in East Anglia (Hopson and Bridge, 1987; Hopson, 1991) revealed green shelly sands at depth, fine to medium grained and well sorted, with a dinoflagellate cyst fauna indicating an open marine environment and probable Antian–Thurnian age. The sands are succeeded by up to 5.8 m of stiff greenish grey silts and silty clays containing wispy stringers and irregular lenses of white fine sand. These resemble the Baventian clays of Chillesford and Easton Bavents and probably were formed in a similar estuarine environment. They are overlain by horizontally bedded brown and orange fine- to medium-grained sands locally interlaminated with thin grey clay seams; these were interpreted as tidally laminated sediments of Baventian to Pastonian age (Hopson, 1991). No gravels were recorded.

The base of the Crag lies at 20 m below OD around Beccles, falling east of Toft Monks [428 948] to between 36 m and 44 m, and finally to 60.1 m below OD in borehole TM 49 SE/2 [4915 9341]. It varies in thickness widely, from 1.5 m in borehole TM 49 SW/12 [4207 9128], where much of the sequence has been cut out by later sands and gravels, to 57.9 m in borehole TM 49 SE/2 in the deep basin to the east. Crag sands crop out on the southern slopes of the Waveney valley either side of Beccles; in the absence of a till at the base of the Corton Formation, the Crag can still be distinguished (although with some difficulty) from the similarly coloured overlying sands of the Corton Formation, since the Crag sands contain iron pan and the Corton Formation sands have a smoother 'feel'.

Kessingland and Pakefield

West (1980) recorded grey and red sandy silt with grey silty clay laminae, of Pre-Pastonian a age, dipping gently northwards and lying directly below the Cromer Forest-bed Formation. These deposits were recorded in boreholes at Pakefield, and exposed at the south end of the coastal section at Kessingland. Blake (1890) and Reid (1890) correlated the latter with the section at the northern end of the Covehithe Cliff. No exposures were found on the beach during the present survey, but Crag sands were proved by augering in Kessingland and along the northern and southern flanks of the valley of the Hundred River.

South Elmham to Wrentham

Around South Elmham the Crag reaches a maximum recorded thickness of 24.7 m in borehole TM 38 SW/9 [3072 8321], and the base falls north-westwards into the Stradbroke Trough, from 2.2 m below OD in borehole TM 28 SW/3 [3419 8242] to 25.6 m below OD in borehole TM 38 SW/7 [3003 8393]. Beneath the till plateau around Brampton and Ilketshall St Lawrence it is difficult, in boreholes, to identify the top of the Crag beneath later arenaceous deposits, but the thickness appears to vary between about 18 and 35 m. This largely comprises sands, although at Sotterley Brickyard, borehole TM 48 SW/1 [4497 8419] records 2.2 m of 'Chillesford Clay' which may correlate with the Baventian clays known farther east. At the top of the local succession, a thick gravel unit crops out along both sides of the valley west of Uggeshall [417 808 to 446 803, 418 807 to 427 803] and also north of Uggeshall [441 808 to 447 812]. The pebbles have a modal diameter of 3 to 6 cm, but cobble-grade gravel underlies finer gravel [442 808] by Town Fen. About 4 m of Crag sands crop out beneath the gravel.

South of Wrentham a bed of silty mudstone up to several metres thick, persistently banded and weathering to a stiff pale buff clay, crops out low in the valley sides at about 5 m above OD. This was formerly dug as a brick clay at Earth Holes [489 805] near Frostenden Corner, and at Cove Bottom [494 798], where the following section was measured:

	Thickness m
Sand, buff, fine- to medium-grained, micaceous, passage base	3.8
Interbedded clayey silt, silty clay and fine-grained sand, micaceous; orange-brown and grey mottled; horizontal bedding, ripple-drift cross-lamination; ferruginous staining on joints; passage base	2.5
Sand, buff, fine- to medium-grained, micaceous	0.9

Covehithe

The Crag thickens eastwards to about 56 m at the coast. Inland it crops out over large areas and augering has demonstrated a coarsening upward sequence: fine- to medium-grained shelly sands at and below 5 m above OD (Plate 7) with lenticles of gravel, mud and sandy mud, pass up into medium-grained sand, then into coarse-grained sands with lenses of gravel. The gravels here are finer-grained and occur as smaller lenses than those farther south, but contain more quartz, quartzite and sandstone, up to around 15 per cent. Several discontinuous clay lenses have been mapped, and demonstrate that the Easton Bavents Clay is not a single bed that could be used as a regional stratigraphic marker.

The cliff sections here have been described by Whitaker (1887), Hey (1967), Long (1974) and West et al. (1980); the sections measured must refer to quite different sites at different stages of coastal erosion. West (1980) described the pollen assemblages in laminated clays from boreholes drilled inland from the Covehithe cliff in 1957, together with a sketch of the cliff section in 1974, which shows the Crag to comprise shelly sands and gravels overlain by a laminated clay unit and then further gravel (Westleton Beds). The following section was measured on the coast [529 821] in October 1977 (West et al., 1980):

	Thickness m
Four fining-upwards units of laminated sand, silt and clay, with desiccation polygons up to 20 cm across on top of each	0.45
Grey-blue sandy silty clay, shelly, bioturbated, desiccation polygon on upper surface	0.80
Grey-blue silt and clay, richly fossiliferous	0.10
Grey-blue laminated clay with shells, closely resembling that at Easton Bavents	0.45
Brown silty clay and sand	0.10

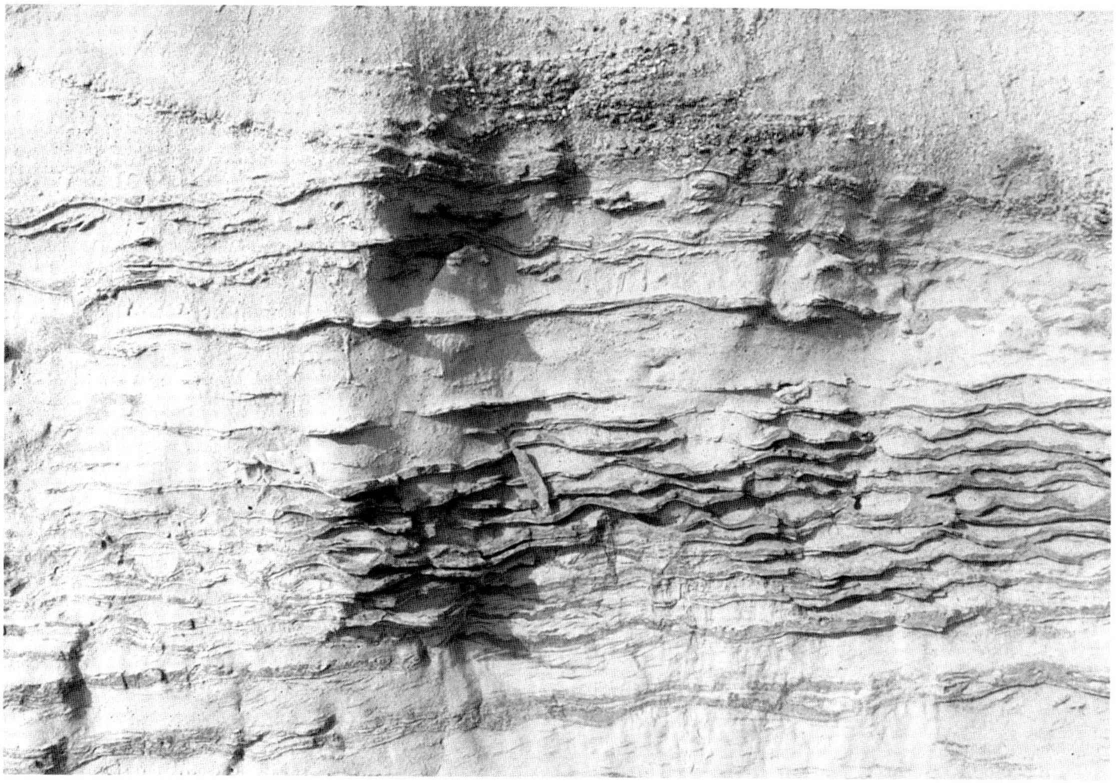

Plate 7 Sands and interbedded thin clays within the Norwich Crag Formation, Covehithe cliffs (GS 566).

Height about 0.5 m

The succession of desiccation polygons shows the uppermost sediments to be intertidal, while foraminifera and molluscs show the laminated clays to be sublittoral to intertidal and imply a boreal to arctic climate.

The current survey revealed several beds of mud in the cliff, including a thick bed forming a platform at the base of the cliff from Covehithe almost to Benacre Ness. Intimately associated with the muds are gravel layers with very variable quartz, quartzite and sandstone contents; these cap the muds and truncate them laterally. The thicker muds, which pass laterally and vertically into sands and gravelly sands, are characterised by lenticular bedding, plant remains, and moulds of bivalves such as *Cerastoderma edule*. The sands are brownish grey to reddish brown and yellowish brown, medium-grained, decalcified, and characterised by trough cross-bedding, ripple-drift cross-lamination, flaser bedding, small-scale channelling, gravel trains and mud drapes. Taken together the strata indicate shallow-marine to intertidal or supratidal environments, in a shifting complex of gravelly beach bars and muddy lagoons, much the same as can be seen along this coast today. The following section [5285 8210] was measured in 1993:

	Thickness m
Sand, yellowish brown, fine-grained, cross-bedded	3.0
Sand, yellowish brown to grey, fine-grained; beds of grey mud and stringers of gravel; cross-bedded and cross-laminated	4.0
Interbedded fine sand and mud	2.0
Mud, mottled grey and brown; stringers and lenses of sand; ripple marks on sandy top surface	1.0+

Easton Bavents

The cliff sections here have been described by Larwood and Martin (1953), Funnell and West (1962), Norton and Beck (1972) and West and Norton (1974). The grey-blue laminated clays found here form the type sediments of the Baventian cold stage (Funnell and West, 1962; Mitchell et al., 1973). In 1987, Mathers and Zalasiewicz (1996) found [516 783] beds of flint gravel up to 0.4 m thick lining channel structures around 2 m deep and 20 m wide, passing up into planar-bedded sands. Up to four such fining-upward units were noted, stacked vertically. Channels of this sort are exposed at intervals for over a kilometre along the low cliffs, and were interpreted as rip-current channels in a near-shore environment.

In 1993 the sections exposed resembled those at Covehithe, with brown-weathering grey muds intercalated with the sands up to about 5 m above OD. The following section was measured [5165 7890]:

	Thickness m
Gravelly sand and sandy gravel, well-rounded flint pebbles and cobbles; channelled base	1.0
Sand, fine- to medium-grained, faintly cross-bedded, with stringers and lenses of gravel	1.5
Gravel, mostly flint, generally well-rounded	0.3
Mud, grey, weathering brown, interbedded with sand, yellow-brown to grey, fine-grained, shelly, cross-laminated	2.0+

Wangford to Southwold

Gravel pits on Wangford Common [around 467 780] (Figure 16a) in 1987 exposed about 10 m of subangular to well-rounded clast-supported medium to coarse flint gravels with a matrix of medium- to coarse-grained quartz sand. The largest flint observed was 16 cm in diameter. The dominant sedimentary structure was large-scale planar cross-stratification, dipping south-eastwards at 6° in the lower 3 to 4 m and 20° in the upper 6 to 7 m, with sporadic up-dip imbrication and thin (less than 0.3 m) sand lenses up to 40 m long. These were interpreted as beachface structures (Mathers and Zalasiewicz, 1996).

Quay Lane Pit, Reydon [483 775] (Figure 16b) in 1987 exposed up to 2 m of gently dipping cross-stratified gravels resembling those seen at Wangford, overlain by up to 4 m of planar and cross-stratified medium- to coarse-grained quartz sand with scattered pebbly beds and lenticular channel-fills. Cross-sets up to 0.5 m thick dipped at 15 to 28° due south. These were overlain by 1.0 to 2.5 m of flint gravel infilling channels, passing up into a similar thickness of planar-bedded medium-grained quartz sand. The succession was interpreted by Mathers and Zalasiewicz (1996) as a transition from gravel-dominated beachface deposits at the base, through sandy shoreface deposits, to an offshore facies dominated by storm-generated sedimentation in rip-current channels.

At Reydon, thick gravels probably form a significant part of the ground now covered by urban development. The cliffs at Southwold are composed of sand, with gravel lenses at the top.

Metfield and Laxfield to Halesworth

The Crag Group does not crop out at the surface, but a thick sequence is preserved in the Stradbroke Trough. The Stradbroke Borehole [232 738], just to the west of the present district, recorded 69.5 m of Red Crag: this is of Ludhamian age between 24.7 m and 69.8 m depth, then Pre-Ludhamian to 94.2 m. Fine-grained grey sands were penetrated down to 31.2 m, then dark blue and grey silt and clay to 33.5 m. Below this the sands were green and grey and became coarser and more shelly, with thin clay layers, claystone nodules and locally cemented sandstone beds. A 1.4 m-thick bed of blue-grey clay occurred at 66.3 m depth, and a 9.9 m bed of clay from 70.9 m to 80.8 m. Below this were further grey silty shelly sands with occasional thin clay layers and claystone nodules. Just above the base were two thin clay layers with coarse shell sand and flints.

The Crag thins eastwards to Laxfield, where boreholes show a sequence of 11 to 15 m of sand, lacking the thick clay beds of the Stradbroke Borehole. Clays are also absent from borehole records in the areas around Linstead Parva and Huntingfield, where the thickest Crag values appear to be respectively 43.9 m in borehole TM 37 NW/7 [3072 7781] and 22.8 m in borehole TM 37 SW/14 [3174 7400]. Boreholes suggest that Westleton Beds gravels are also absent from the sequence in these areas, but a thick gravel bed crops out in the valley on the north and east sides of Halesworth, and gravels also appear in the village of Bramfield. It is notable that in the area south of Halesworth, where the gravels are absent, the Crag is recorded up to 29.8 m thick in borehole TM 37 SE/6 [3904 7074], but in the first borehole at Bramfield to record gravel at the top of the sequence (TM 37 SE/15 [3910 7475]) only 8.5 m of Crag sands are recorded below the gravel, supporting the suggestion of an unconformity below the Norwich Crag.

Holton, Thorington, Dunwich

In the area of Holton and Wenhaston, boreholes show that the base of the Crag lies at 4 to 11 m below OD, and the thickest record is 31.7 m (TM 47NW/6 [414 790]). Westleton Beds gravels are widely represented at the top of the sequence, but are known as low as 18 m above the base of the Crag (borehole TM 47NW/1 [404 760]). Generally the gravels occur as lenticles up to a few hundreds of metres long, but a much larger continuous body occurs along the north side of the Blyth valley [402 774 to 438 765]. Before the erosion of the valley through Holton this might have been continuous with the gravels underlying the northern and eastern parts of Halesworth. The overall morphology of the gravel body indicates a southerly dip of about 1°, but boreholes demonstrate that it does not reappear on the south side of the Blyth valley, and in the large abandoned gravel pit at Holton [405 773], Hey (1967) recorded dips of 3 to 10° to the south-east. The section recorded by Hey (1967) was 12.2 m high and 315 m long, with cross-bedded gravels passing laterally into sands. His analysis of the 16–32 mm fraction yielded 78 per cent rounded flint, 20 per cent subangular and angular flint, and 1 per cent each of quartz and quartzite or sandstone. The following section was measured in 1992 at the highest part of the rear face of the quarry:

	Thickness m
CRAG	
Sand, pale to medium brown, medium-grained, with low-angle cross-bedding; passage base	0.5
Gravel, rounded flints up to 10 cm long, about 10% quartz and quartzite; matrix-supported in medium- to coarse-grained pale to medium brown sand	0.3
Sand, buff to pale brown, coarse-grained; iron staining picking out cross-bedding; scattered flint pebbles to 2 cm long, the larger pebbles being of high sphericity and roundness	1.6
Sand, buff and orange, coarse-grained; pebbles up to 5 cm long isolated and in layers; more pebbly and deeper orange below; passage base	c.2.0
Gravel, rounded flints up to 10 cm long, rare quartzites; matrix-supported in buff sand; layers of sand up to 10 cm thick; no cross-bedding visible	c.1.5+

Hey (1980) considered that gravels of the Kesgrave Group were present overlying the Crag, but none was noted during the present survey. No stratigraphical break could be detected in the section detailed above, and none is implied by heavy mineral analyses from three levels within the section (Hallsworth, 1994). At the old gravel pit [423 770] near Blyford Church, Whitaker (1887) recorded tabular masses of grey clay at the base of the gravels.

Fine-grained sands outcrop widely on both sides of the Blyth valley, underlying the gravel body on the north side; shelly sands are found within the weathered zone in a pit [442 754] near Union Farm, yielding disarticulated bivalves (strongly and weakly ribbed), turritellids and other gastropods. Fossil lists are given by Prestwich (1871a p.344) and Whitaker (1887 p.81). At the Blyth River Gravel Pit [410 767], Crag sands are worked for moulding sand and soft building sand. The sands worked lie wholly within the weathered zone and are fine to medium-grained, with cross-bedded units up to 20 cm thick and much iron pan. Beds of gravel (Plate 5) up to 70 cm thick at one point [4119 7672] contain rounded flints up to 10 cm long but very little quartz or quartzite.

From Thorington to Darsham and Dunwich the Crag is up to around 40 m thick, and crops out widely in the east along the valley of the Dunwich River and on the interfluve of Dunwich Forest. Gravels occur throughout the area, and in Dunwich Forest the area of gravel may be underestimated because of the difficulty of distinguishing Crag gravel from glaciofluvial deposits in continuous woodland with no field brash. Several gravel bodies have been mapped in the area east of Bramfield, but these do not extend as far to the south-west as Darsham.

Clays have been found in the sands underlying the gravels at Wenhaston [4230 7479] (Whitaker, 1887 p.9) and in auger holes [431 711] north-west of Red House Farm. A borehole for water (TM 47 SW/8 [4194 7264]) at Park Farm recorded two beds of gravel separated by a 2.5 m band of clay with sand partings; the lower gravel was underlain by at least a further 22 m of sand. The same succession was observed in 1992 in the working face of the adjacent gravel pit [4238 7281]:

	Thickness m
GLACIOFLUVIAL DEPOSITS	
CRAG	
Sand, brown, buff, and grey; ripple drift lamination dipping north, rounded flints scattered and in layers, little quartz and quartzite	5.0
Gravel, matrix-supported, rounded flints up to 10 cm long	0.6
Sand, brown, medium-grained, micaceous	0.2
Gravel, as above	0.1
Clay, grey with brown and black beds; silty, well laminated	1.0
Gravel, as above	0.15
Sand, as above	0.3
Gravel, matrix-supported, rounded high-sphericity chatter-marked flints in coarse sand, few quartz and quartzite pebbles, layers of medium-grained gritty sand up to 30 cm thick with cross-bedding dipping north; some gravel layers finer than others; flints up to 20 cm long	c. 2.3+

The suggestion was considered that the 5 m sand unit below the glaciofluvial deposits could belong to the Kesgrave Group or Bytham Formation. However, there was no clear break between this bed and the underlying gravel, and the heavy mineral suites from this bed and from clay and sand beds lower in the sequence are similar (Hallsworth, 1994). This bed is likely to represent a sandy shoreface facies of the Norwich Crag like that which overlies the gravels at Reydon, where the cross-stratification in the sands similarly has a unimodal dip direction.

Peasenhall to Westleton

Buff and orange sands are seen at outcrop in the valley of the River Yox from Peasenhall to Yoxford. Borehole records mention clays and gravels within the succession but no clear stratigraphy is apparent. Downstream of Middleton the Crag crops out widely on either side of the River Yox and across Westleton Walks and Dunwich Heath to the north. The first gravel outcrop seen at surface is an overgrown pit [420 691] south of Darsham Church, and large areas have been mapped east of Westleton, but the area of gravel shown hereabouts on the published 1:50 000 sheet may be exaggerated because of the problems of mapping out gravel from sand in dry, well-drained heathland.

The gravels have been widely worked south and south-east of Westleton [4440 6885, 4430 6860, 4450 6870], but all the workings are now disused and overgrown. Prestwich noted (1871b p.461-2) 30 to 40 feet (9.1 to 12.1 m) of 'stratified beds of well-rounded flint-pebbles imbedded in white sand, and with two or three subordinate beds of light-coloured clay', while Hey (1967) described two large pits [443 686, 445 687] in which clast-supported gravels passed up into sand with scattered pebbles, with consistent 10° dips to the south-east. Pebble counts for the 16–32 mm fraction at the latter pit (Hey, 1967) revealed 54 per cent rounded flint, 42 per cent subangular and angular flint, 3 per cent quartz and 1 per cent quartzite and sandstone.

Framlingham, Saxmundham

The extent of the Crag is difficult to ascertain because of the poor quality of available borehole logs and the paucity of exposures. Previous small-scale geological maps had shown the Crag to be absent around Framlingham, but reappraisal of well records suggests that it is present except for a small area at Dennington [27 66]. Around Saxmundham the Crag is better known from boreholes, thinning from 30 m in the west to about 12 m in the east and cropping out along the sides of the valley of the rivers Alde and Fromus. Boreholes record seams of clay and silt within the sands, and a bed of flints is commonly recorded at the base of the Crag.

Easton to Snape

In the south-west of the district, the Crag is around 25 m thick. The higher parts of the succession crop out locally in the flanks of the valley of the River Deben [30 56], and in the valley below Framlingham downstream from Moat Hall Pit [305 5972]. Here 2.5 m of cross-bedded yellow to dark orange-brown sands were seen overlain by quartz- and quartzite-rich gravels ascribed to the Kesgrave Group.

In the Snape area the Crag Group is also around 25 m thick, and can be subdivided into Red Crag of Pre-Ludhamian age and Norwich Crag of ?Bramertonian age. The Red Crag underlies the whole area but only crops out in the lower slopes of the Alde Valley south-west of Snape [385 577, 379 572, 388 578] and in a tributary valley [358 583]; it comprises medium- and coarse-grained poorly sorted shelly sands, rich in comminuted molluscan debris, with rare small flint, vein quartz and phosphate pebbles but little silt- and clay-grade material. The overlying Norwich Crag or Chillesford Sand (Zalasiewicz and Mathers, 1985) is 10 to 15 m thick and crops out widely in the sides of the major valleys. It comprises unfossiliferous fine- to medium-grained well-sorted sand, with scattered silt and clay laminae, horizontal bedding and burrows; coarser layers are present near the base, making the mapping of the boundary with the Red Crag difficult.

Aldeburgh to Sizewell

Around Aldeburgh and towards Orford to the south of the district, Mathers and Zalasiewicz (1985) investigated the Crag deposits using extendable augers, trial pits, rotary drilling, shell and auger boreholes and electrical conductivity measurement. They found the Red Crag to be banked up against the Coralline Crag ridge, and the distribution of the deposits to be almost mutually exclusive. The Red Crag, here up to 25 m thick, comprised a basal bed of phosphatic nodules, then medium- and coarse-grained shelly sands with large-scale cross-stratification passing up into shelly sands with small-scale stratification and horizontal bedding. The Norwich Crag locally oversteps the Red Crag to rest on Coralline Crag; it was distinguished from the Red Crag by its finer grain size and higher degree of sorting, although the junction was hard to define because of the presence of coarse beds within the basal Norwich Crag.

The outcrop of the Chillesford Clay was determined by conductivity mapping (Mathers and Zalasiewicz, 1985; Cornwell, 1985; Zalasiewicz et al., 1985), the clay having a higher conductivity than the sands. It occupies a south-west-trending tract about 3 km wide, from north of Aldeburgh to around Butley [369 511] in the Woodbridge district, with three outliers north of the River Alde. It passes laterally into the upper part of the Chillesford Sand, and comprises up to 5 m of grey, silty, variably micaceous clays with fine sand and silt laminae up to 5 mm

thick, and rare molluscan shell debris. The deposit has been extensively worked for brickmaking, although only one pit remains [452 572].

Zalasiewicz et al. (1988) reported on six rotary-cored boreholes spaced at 0.5 to 1.0 km intervals between Aldeburgh and Sizewell. These demonstrated a thick Crag sequence within a deep sharply bounded basin, which the authors considered to be of erosional rather than tectonic origin. Within the basin, towards Sizewell, Norwich Crag of Bramertonian age rests unconformably on the Sizewell and Thorpeness members of the Red Crag, which rests upon the Thames Group at 40 to 50 m below OD; while on the edge of the basin, nearer to Aldeburgh, thin Norwich Crag rests on Coralline Crag, which rests upon Thames Group at 15 m below OD. The Crag Group thus increases in thickness from about 20 m at Aldeburgh to about 60 m at Sizewell. The Red and Norwich crags were thickest (respectively 31.4 m and 16.6 m) in borehole TM 45 NE/9 [4517 5925] north-west of Aldeburgh (Figure 17). Surface exposures of Norwich Crag in the vicinity of Sizewell have been assigned to the Pastonian (West and Norton, 1974) or Bramertonian (Funnell et al., 1979). Micropalaeontological studies of borehole samples obtained during site investigations for the power stations identified ?Baventian, ?Thurnian, ?Antian and Pre-Ludhamian stages (West and Norton, 1974; Funnell, 1983b).

CROMER FOREST-BED FORMATION

Nomenclature and distribution

The term Cromer Forest-bed Series was given by Reid (1882) to a sequence of marine, brackish and freshwater sediments deposited in a wide embayment in the coastal region of northern and north-eastern East Anglia, and exposed discontinuously along the coast from Weybourne in Norfolk to Kessingland in Suffolk. The maximum thickness is 8 m. The Cromer Forest-bed Formation (Funnell and West, 1977) was included in the Kesgrave Group by Arthurton et al. (1994). This practice is not adopted here, although it is likely that the formation is at least in part the downstream equivalent of the Kesgrave and Bytham fluvial formations. Offshore it has been correlated with the delta-top Yarmouth Roads Formation (Bowen et al., 1986; Arthurton et al., 1994); the latter has been dated as Pastonian to Beestonian (Funnell, 1987) or Cromerian Complex (Cameron et al., 1992). The lower part of the Cromer Forest-bed Formation may correlate with the Markham's Hole, Outer Silver Pit and Aurora formations offshore (Table 5). Within the present district, strata assigned to the Cromer Forest-bed Formation have yielded dates between Pastonian II and Cromerian IIIa (West, 1980).

The deposits crop out at the base of the sea cliffs between Pakefield and Kessingland (Blake, 1884a, b; West, 1980); they are not believed to have extended much farther south and are absent from the Covehithe sections (Funnell and West, 1977; West, 1980). South of the Waveney valley they do not extend far inland, since they rest on a bench bevelled into the Crag Group, which crops out at higher levels in the village of Kessingland [535 861, 532 864], at Black Street [518 870] and at Gisleham [513 886]. North of Pakefield they are not seen in the cliffs again until Corton (Arthurton et al., 1994); in the intervening area they may have been removed by erosion beneath the Corton Formation. Within the valley of the Waveney, inland of Lowestoft, they are known from boreholes in the 'Corton Embayment' (Hopson and Bridge, 1987), where they rest on Crag at about OD. In this district the embayment extends inland to Waterheath [44 94], west of which it is

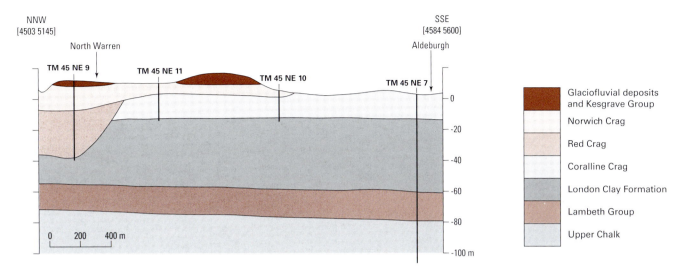

Figure 17   Geological cross-section from Aldeburgh to North Warren (after Mathers and Zalasiewicz, 1988).

bevelled into the Toft Monks ridge where the Bytham Sands and Gravels rest upon Crag. To the south the embayment does not extend as far as Barnby [47 89], because Crag crops out on the flanks of the Waveney valley there.

Stratigraphy and sedimentology

The stratigraphy was first elucidated on the coast between Pakefield and Kessingland by Blake (1884a, b; 1890) and Reid (1890). Blake's descriptions may be summarised as follows:

	Thickness m
'Freshwater Beds': laminated black, brown and grey clay, ferruginous gravel, brown and orange pebbly sand; includes at the base a flattened oak tree trunk 22 ft (6.7 m) in length	not stated
'Unio-bed': ferruginous flint gravel with freshwater shells	0–0.15
Peat and silt, with seeds, matted reeds and compressed wood	0.15–0.3
'Rootlet-bed' ('Forest-bed'): unstratified homogenous greenish grey clay with calcareous concretions ('race'), small flints and mammalian remains; rootlets up to 1.8 m long, 4 cm thick	1.2–3.0
'Chillesford Beds': alternations of micaceous grey clay and white and brown sand; unfossiliferous; passage base	3.4–3.7
'Gravel with mammalian remains': stratified sand and gravel, ferruginous, indurated, with grey clay seams and bones of elephant, rhinoceras, horse and deer	not stated

Although the succession given above clearly accords with the horizontal section published by Blake (1884a), he later reversed the order of superposition of the 'Chillesford Beds' and the underlying gravel under pressure from his colleagues (Blake, 1890), since the 'Chillesford Beds' were assumed to be a part of the Crag, while the gravels contained mammalian remains appertaining to the Cromer Forest-bed.

The deposits were ascribed to the Cromerian stage by West

and Wilson (1966). West (1980) remeasured the sequence and interpreted it as a complex of marine and freshwater strata deposited on a low-lying coastal plain as contemporary sea levels varied during the period from Pre-Pastonian to Cromerian. His succession is as follows, with his interpretations with regard to age, environment and climate:

	Thickness m
(Bed i: sands of the Corton Formation)	—
Bed h: laminated brown and red silty clays, tidal; transitional base	0.3–0.8
Bed g: brown silty mud with laminae richer in silt; basal part sandy with organic debris including twigs, leaves; Blake's 'Freshwater Bed'; freshwater to brackish, Cromerian IIIa age	0.8–1.6
Bed f: red ferruginous sand and gravel; includes Blake's Unio-bed at the southern margin of a channel cut into Beds d and e; freshwater	up to 0.25
Bed e: peat with wood and seeds; Cromerian II	up to 0.30
Bed d: 'Rootlet Bed': stiff brown or blue clay with rootlets and 'race' overlying brown clayey silt with sand beds; Cromerian II	up to 3.5
Bed c: brownish red sand with pale flint chips, small rounded flints more abundant at the base; Reid (1890, p.135) noted marine shells	2.0
Bed b: grey and brown laminated silty clay with horizons of ferruginous or grey silty sand and flattened wood; temperate climate, shallow marine conditions; Pastonian II age	0.7
(Bed a: grey and red sandy silt with grey silty clay laminae; marine; Pre-Pastonian a, Crag Group)	—

Beds f, g and h are interpreted as temperate-climate strata infilling a channel cut through the Rootlet Bed. Elsewhere in his section Blake (1884a) showed Bed e to be overlain by laminated clay and sand; West (1980) failed to excavate this bed but considered by correlation with Corton that Bed e and this overlying bed are of Cromerian II and IIIb ages respectively.

Boreholes in the Corton Embayment (Hopson and Bridge, 1987; Hopson, 1991) record a tripartite sequence in which up to

2 m of waxy bluish grey silty clay with small 'race' nodules, correlated with the Rootlet Bed of the coast, are underlain and overlain by flat-lying units up to 5 m thick of interlaminated sand, clay and gravel. Both of the the gravelly units show evidence of marine and fluviatile deposition. The lower unit is channelled into the underlying Crag at about OD; the clays at the base are correlated with the marine Bed b of West (1980) and represent an early transgression, while the associated gravels and the succeeding rootlet bed (Bed d equivalent) represent a return to freshwater conditions. Within the upper gravelly unit, dinoflagellate cysts of inshore or semi-lagoonal affinity (Harland, 1984) were recovered from tidally laminated sediments indicating a second marine transgression to at least 7 m above OD. This will correlate with Bed g (Cromerian IIIa) which West (1980) recorded at 2.3 to 4.0 m above OD on the coast. There is no evidence that either marine transgression overtopped the Toft Monks ridge. Both the gravelly units are poorly sorted and heavily iron-stained, with clasts in the +8 -16 mm fraction dominated by flint (73 per cent), vein-quartz (13 per cent) and quartzite (9 per cent). This places them within the range of flint:quartz:quartzite ratios recorded from the Bytham Sands and Gravels (Table 6, Figure 19), but with a relatively low percentage of quartzite.

Details

Coast section, Kessingland to Pakefield.

When Blake (1884a, b; 1890) examined this coastline, erosion by the sea was proceeding at a rate estimated at 'considerably more than a yard' per year, exposing a much more complete section than has been seen since. He showed (1884a) the Rootlet Bed and underlying clays (which he correlated with the Chillesford Clay) to have been exposed over most of the distance from Kessingland [northing 8590] to a point [northing 8835] (eastings not known because of subsequent cliff retreat) about 320 m south of Crazy Mary's Hole. For the next 230 m northwards, the Rootlet Bed was cut out below a channel filled with gravel and sand, and thence the clays underlying the Rootlet Bed were again exposed to a point [northing 8968] about 1000 m north of Crazy Mary's Hole.

West (1980) found the section to have been obscured by talus as a result of beach aggradation south of Lowestoft. He found one natural section, at Kessingland, which exposed the Rootlet Bed, including an organic horizon, resting on brown sand with clay seams which he regarded as Crag (Bed a). At Pakefield he made eight excavations and sunk two boreholes, sited to investigate Blake's Rootlet Bed, the channel filled with organic sediments, and the underlying clays which Blake had correlated with the Chillesford Clay.

At the time of the present survey (1993), well-laminated fine- to medium-grained buff sand and grey clay with ferruginous sands were exposed beneath the Corton Formation on the south side of Crazy Mary's Hole [5366 8867]; by correlation with Blake's (1884) section this should be lower than the Rootlet Bed, and is thus probably Bed c of West (1980). Farther south [5363 8836 to 5365 8848] at beach level, the following section was observed. It is believed to represent a channel infill equivalent to Beds g and f of West (1980), although rather coarser-grained:

	Thickness m
Sand, dark greyish brown, medium-grained, micaceous, clayey, flat laminated; seams of grey clay; layers and nodules of iron pan; passage base	1.0
Clayey sand, black, fine- to medium-grained, with thin clay seams; micaceous, flat laminated, peaty and organic-rich	1.0

Waveney valley

Four boreholes prove strata of the Cromer Forest-bed Formation; TM49SW/135 at Bull's Green [4179 9423] proved 0.5 m of brown and grey sticky silty clay with 'race' nodules, lithologically identical to the Rootlet Bed of the coast (West, 1980) but some 13 m higher. It rests upon Bytham Sands and Gravels of the Toft Monks Ridge. Boreholes TM49SW/137, TM49SE/8 and 9 [4412 9440, 4644 9326, 4831 9303] record strata laid down within the Corton Embayment, comprising quartz- and quartzite-bearing sandy gravels, pebbly sands, and sand with thin silty clays. The base of these deposits is at 1.0 to 2.8 m below OD and their top is at 3.2 to 7.4 m above OD. In adjacent areas to the north (Bridge, 1993) the same strata are divided by the Rootlet Bed into upper and lower units.

Along the lower flanks of the valley and on slopes overlooking Oulton Broad, sand and gravel which may belong to the Cromer Forest-bed crop out between OD and 5 m above OD. The gravel fraction comprises mostly subangular to rounded flint but also includes quartz and quartzite. A pit [5125 9200] revealed 0.5 m of gravel, resting on Crag and overlain by sands believed to belong to the Aldeby Sands and Gravels. Many boreholes in and around Lowestoft record gravels within sand at this topographic level, but none can be confirmed as Cromer Forest-bed Formation.

KESGRAVE GROUP

Nomenclature and distribution

Throughout much of Essex and southern Suffolk a suite of quartz- and quartzite-bearing sands, gravels and silts is present beneath Anglian glacial deposits. These were mapped as 'Glacial' on the original Geological Survey maps, although Prestwich (1871b, 1890) recognised their preglacial origin. However, he confused them with the flint gravels in the Crag Group, considering them to be marine and referring to them as Westleton Beds. The term Kesgrave Sands and Gravels was coined by Rose et al. (1976), with a type area in the Gipping Valley. They were later renamed Kesgrave Formation (Hey, 1980; Rose et al., 1985), and shown to be fluvial in origin, representing early terraces of the River Thames which, prior to the Anglian glaciation, followed a more northerly course than it does now. As such they have been variously subdivided (Hey, 1980; Green, McGregor and Evans, 1982; Bridgland, 1988). Most recently they have been restyled the Kesgrave Group (Whiteman, 1992; Whiteman and Rose, 1992) and divided into the Sudbury Formation (older) and Colchester Formation (younger), these in turn being divided into members representing individual terraces of the Thames. Eight out of the ten members have been recognised in East Anglia. The members are distinguished on the basis of terrace elevation and clast content; in descending order they are the Bushett Farm Gravel, Stebbing Gravel, Bures Gravel and Moreton Gravel (in the Sudbury Formation), and the Waldringfield Gravel Ardleigh Gravel, Wivenhoe Gravel and Lower St Osyth Gravel (in the Colchester Formation).

Although most workers have concentrated their researches in Essex and southern Suffolk as far as Ipswich, occurrences of quartz- and quartzite-rich gravels are known farther north in Suffolk and in Norfolk (Hey

Table 6 Clast lithological analysis of samples from the Bytham Sands and Gravels, with comparative analyses from the Corton and Cromer Forest-bed Formations.

Site borehole number	National Grid Reference	No. of samples analysed	% angular flint	% rounded flint	% vein quartz	% quartzite	% chalk	% other limestone	% igneous & meta-morphic	others	
TM28NE-17	2916 8550	6	43	13	16	26	0	0	trace	2	
30	2758 8810	1	33	8	22	34	0	0	1	2	
31	2894 8764	3	47	19	11	18	0	0	0	5	
TM28SE- 26	2596 8356	1	57	7	19	16	1	0	0	0	
28	2517 8066	1	60	27	4	6	trace	0	0	3	
31	2630 8048	1	71	7	9	6	0	0	1	7	
31	2630 8048	1	31	34	10	16	0	0	0	9	
32	2741 8454	4	39	24	16	17	trace	0	0	4	
37	2708 8026	3	35	41	14	19	trace	0	0	1	'Kesgrave' and 'Pebbly Series' samples from Auton et al. (1985)
40	2857 8391	8	50	14	16	17	trace	trace	trace	3	
41	2838 8287	1	38	26	14	18	0	0	0	4	
42	2864 8163	3	54	28	8	7	trace	0	0	3	
53	2741 8047	1	46	31	5	16	0	0	0	2	
TM27NE- 5	2554 7995	1	44	39	8	8	0	0	0	1	
5	2554 7995	1	45	10	22	23	0	0	0	0	
6	2577 7870	2	44	30	10	9	0	0	4	3	
7	2502 7606	2	41	14	24	18	0	0	1	2	
10	2735 7630	2	38	22	26	12	0	0	1	3	
15	2876 7894	1	44	29	11	16	0	0	0	0	
TM39SW-36	3105 9407	2	28	27	18	25	0	0	3	2	
51	3389 9193	1	50	22	8	13	0	0	1	6	
51	3389 9193	1	42	30	13	7	0	0	2	6	
TM38NW-35	3169 8931	1	38	25	16	16	0	0	trace	5	
43	3351 8803	6	44	18	16	15	trace	0	1	6	
Toft Monks Ridge		12	19	47	10	20	trace	1	trace	3	
Corton Formation (gravel)		68	54	17	12	10	trace	2	1	4	Hopson and Bridge (1985)
Corton Formation (till)		10	48	15	16	11	trace	3	1	6	
Cromer Forest-bed (gravel)		46	49	24	13	9	0	2	trace	3	

and Brenchley, 1977; Hey, 1980; Auton et al., 1985; Green and McGregor, 1990). This has led to a belief that the sediments of the Kesgrave Group were deposited over a very large area of Suffolk and Norfolk by a proto-Thames flowing north-east and north (Whiteman, 1992, fig. 2; Arthurton et al., 1994, fig. 29), moving its thalweg successively south-eastward as it cut lower terraces (Whiteman, 1992, fig. 6). The whole of the Saxmundham–Lowestoft district would in that case lie within the area covered by the Group. However, in the current survey quartz-rich gravels were recorded in two separate areas; those in the southern part of the district are part of the Kesgrave Group, but those in the north, termed Bytham Sands and Gravels (Bateman and Rose, 1994; Rose, 1994) were probably deposited by a separate river system. It is now thought likely, therefore, that there were two river systems flowing across Suffolk (Figure 18), the proto-Thames in the south and the Bytham River (Rose, 1994) in the north (Hopson and Bridge, 1987). The deposits of the latter river are described separately.

Within the present district it has not proved possible to subdivide the Kesgrave Group as has been done in the Thames Valley, although the palaeogeographical recon-struction of Whiteman (1992, fig. 6) suggests that the highest five members are, or have been, represented within the district. Also, it has not proved possible to show the extent of the Kesgrave Group on the maps, because of the difficulty of recognising the group at outcrop beneath glaciofluvial sands and gravels of the Lowestoft Till Formation. The glaciofluvial deposits normally include cobble-grade clasts of angular flint and chalk (except where they have been decalcified by surface weathering) but these clast types are not represented in the Kesgrave Group. However, the Kesgrave Group includes no litholo-gies which do not also occur in the glaciofluvial deposits. Moreover, the relative proportions of flint, quartz and quartzite in the two deposits are similar, suggesting that the later deposit may incorporate much material locally reworked from the earlier. The unconsolidated glacio-fluvial sands and gravels are prone to wash downhill over the Kesgrave Group outcrop, hence the outcrop of the Kesgrave Group cannot be identified either by examina-tion of the soil brash or by feature-mapping. For these reasons the Kesgrave Group and the glaciofluvial sands and gravels have been mapped as a single unit and included in the Lowestoft Till Formation.

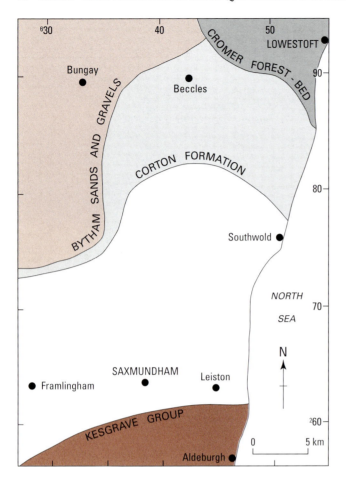

Figure 18 Approximate limits of the Kesgrave Group, Bytham Sands and Gravels, Cromer Forest-bed Formation and Corton Formation in the district.

It is possible that the Kesgrave Group may overlie the Crag in this district as far north as Leiston, though the northernmost exposure known [at 4620 5990] is only tentatively identified as Kesgrave Group. Farther north the Group is absent from the valley of the River Yox and Minsmere River, where the fluvioglacial sands and gravels are more restricted; in many places Lowestoft till can be seen to overlie Crag with no intervening Kesgrave Group.

Stratigraphy and sedimentation

The gravels and sands of the Kesgrave Group are considered to represent the sediments of a braided river, formed during successive periglacial periods (Rose and Allen, 1977). The Group is considered to be the upstream, fluviatile equivalent of the Yarmouth Roads Formation (Bowen et al., 1986, Arthurton et al., 1994), a seismostratigraphical unit recognised offshore and interpreted as a delta-top complex (Balson and Cameron, 1985; Cameron et al., 1992).

The sediments largely comprise unfossiliferous 'clean' medium- to coarse-grained sands, interbedded with pebble-grade gravels and lenses of silt and clay. Rose and Allen (1977) recorded sedimentary structures including

channel fills, planar beds and large- and small-scale cross-bedding with a dominant foreset dip to between 010 and 100°, implying a north-eastward direction of river flow. The gravels are not as coarse as those within the Crag or the glaciofluvial deposits of the Lowestoft Till Formation. Except where the top metre of sediment is reddened by palaeosol development, the sands vary from white through buff to orange according to their iron content. White mica and black iron oxide occur, but carbonaceous material is not recorded, while thin seams of greenish or grey silt and silty clay are present. The gravel clasts are mostly rounded to subangular flint, but the most characteristic feature of the deposit is its content (up to 27 per cent) of rounded pebbles of white vein quartz and purple, red or white quartzite. Clasts of igneous and metamorphic rocks are present in small quantities, with sponge-spicular chert which Bridgland (1988) considers to be derived from the Lower Greensand of Kent and Surrey.

Hey (1980) considered that the quartzite pebbles are all 'Bunters' derived from the Triassic rocks of the Midlands, the colourless varieties having been bleached of their original iron oxide colours. Some of the quartzites retain a pink or purple core beneath a colourless exterior, and in thin sections coloured and colourless varieties are indistinguishable (Hey, 1976, pp.75–76). The igneous rocks almost all belong to a single suite of fine-grained acidic lavas and vitric tuffs, which Hey and Brenchley (1977) concluded were derived from the Ordovician rocks of North Wales.

In southern Suffolk deposits of the Kesgrave Group are overlain by a complex of soil profiles revealing iron enrichment and rubification, gleying, clay illuviation, and periglacial disruptions such as involutions and ice-wedge casts (Kemp, 1985, 1987a). At Ipswich Airport this complex overlies the Waldringfield Gravel, the earliest and highest terrace of the Colchester Formation (Kemp, 1987b). The oldest member of the soils complex here is the Valley Farm Rubified Sol Lessivé, which is interpreted as having formed during the Cromerian, the last temperate stage before the Anglian. The superimposed Barham Arctic Structure Soil formed in periglacial conditions early in the Anglian (Rose and Allen, 1977; Rose et al., 1985). On the basis of these ages, Rose et al. (1976) placed the Kesgrave Sands and Gravels in the Beestonian, the cold stage immediately before the Cromerian, but it has become apparent that the formation of at least ten terraces of the Thames would take a much longer time, extending well back into the Early Pleistocene (Whiteman and Rose, 1992). Thus, where it is formed on the higher, earlier members of the Kesgrave Group the Valley Farm Soil might span the Pastonian and Cromerian temperate phases and the intervening Beestonian periglacial phase (Kemp, 1985). The Valley Farm and Barham soils have not been recorded in association with the Kesgrave Group in the present district, although periglacial involutions noted at the top of the Crag at Thorington may date from the same early Anglian period as the Barham Arctic Structure Soil, and rubified palaeosols are known at the top of the Bytham Sands and Gravels and the Crag Group in the north of the district (Auton et al., 1985).

Bowen et al. (1986) dated the Kesgrave Group as Pre-Pastonian to Beestonian. Whiteman and Rose (1992) suggested that the Sudbury Formation, which is believed to have been formed at a time of glaciation in Wales and the Midlands, may be Pre-Pastonian to Beestonian, while the Colchester Formation is of 'Cromerian Complex' age, the total Kesgrave Group spanning some 1.2 million years. Certainly, if the Cromerian age of the Valley Farm Sol Lessivé is correct then that would be the youngest possible date for the higher (older) members of the Group, although Bridgland (1988) suggests that formation of the lower (younger) members may have extended into the Anglian. Also, since it is reasonable to assume the existence of a proto-Thames river during the period in which the Crag was forming, the highest terraces might be contemporaneous with the Crag.

Details

Blaxhall–Snape

The Kesgrave Group is believed to crop out widely on the north and south flanks of the valley of the River Alde, but its outcrop cannot be separated from that of the overlying glaciofluvial sands and gravels which wash down over it. The deposits are 3 to 6 m thick, possibly thicker in localised scourhollows, with a base at about 15 m above OD. The only section is that recorded in a pit [3788 5700] on Blaxhall Common:

	Thickness m
Soil and wash	0.50
KESGRAVE GROUP	
Sand, yellow-grey, medium- to coarse-grained, poorly sorted, gravelly, pebbles up to 20 mm diameter of subangular flint, rounded and subrounded quartz and quartzite; laminae dip gently north-east; sharp base with gravel lag	0.40
NORWICH CRAG	
Sand, fine-grained	1.00

Holes drilled with an extendable auger [3805 5580, 3904 5581] just south of the district revealed up to 3.3 m of Kesgrave sands, not bottomed, with a few scattered pebbles of flint and quartz.

Snape Warren–Friston–Aldeburgh

The Kesgrave Group underlies interfluves which form broad plateaux at heights of 12 to 20 m above OD, but on the map cannot be distinguished from the overlying glaciofluvial sands and gravels. The base of the Group, which together with the glaciofluvial deposits is about 8 m thick, is believed to drop gently northward. Soil brash implies that the deposits are largely sands with only restricted development of gravels. The thickest record was at a pit at Aldeburgh [4523 5724]:

	Thickness m
KESGRAVE GROUP	
Sand, fine- to coarse-grained, moderately sorted; pebbly, dominantly flint and quartz pebbles up to 30 mm diameter; undulatory base	3.4
NORWICH CRAG (Chillesford Clay)	
Clay, silty, grey-brown, mottled	1.6

Thorpeness–Aldringham–Leiston

A small exposure at Thorpeness [4620 5990] revealed 0.6 m of gravel, with flint and quartz pebbles up to 50 mm in diameter and cross-bedding dipping at 10° towards 024°. It is uncertain whether this is Kesgrave Group or glaciofluvial sands and gravels, although the direction of dip would accord with the former. The site lies at the south-eastern end of a broad expanse of glaciofluvial deposits extending to Aldringham and Leiston, and which may or may not be underlain by Kesgrave Group; exposures are lacking.

BYTHAM SANDS AND GRAVELS

Nomenclature and distribution

The Bytham Sands and Gravels (Bateman and Rose, 1994; Rose 1994) are a package of quartz- and quartzite-rich fluvial sediments which can be traced (though with gaps) from Warwickshire to the present district (Rose, 1987) and which appear to have been deposited over a considerable period of time by a single river system, the 'Bytham river'. They include the Ingham Sand and Gravel of Clarke and Auton (1982), a quartzite-rich deposit which rests on Chalk at 8.4 to 20.9 m above OD around Redgrave in the Upper Waveney valley. Lewis (1993) used the term Ingham Formation to cover all the sediments of the Bytham river in eastern Suffolk, so making it a synonym of the Bytham Sands and Gravels. He divided the formation into four members corresponding to successive river terraces; from older (higher) to younger these are the Seven Hills Gravel, Ingham Gravel, Knettishall Gravel and Timworth Gravel. However, in this memoir the term Bytham Sands and Gravels is used in the sense of Bateman and Rose (1994), and Rose (1994) to include all the deposits of the Bytham river, since the term Ingham Sand and Gravel was restricted by Clarke and Auton (1982) and by Mathers et al. (1993) to deposits with a particular ratio of flint, quartz and quartzite, and was not intended to include all of the post-Crag pre-Anglian gravels in the area.

The existence of a suite of pre-Anglian quartzose gravels associated with the valley of the Waveney and physically separated from the Kesgrave Group to the south was recognised by Hopson and Bridge (1987), who examined regional variations in the main constituents (flint, quartz and quartzite) of such gravels. They demonstrated that the high proportion of quartzite to vein-quartz in the deposits of the Waveney valley (Diss and Redgrave areas) was significant and distinguished those deposits from the Kesgrave material of the area between Stansted, Sudbury and Nayland. Clarke and Auton (1982) and Hey (1980) had attempted to explain this anomaly by suggesting that additional quartzite was added to the northward-flowing Kesgrave (ancestral Thames) river by a tributary which crossed the Chalk through a gap between Bury St Edmunds and Thetford. However, the ratio of quartzite to quartz in the Bytham Sands and Gravels of the Toft Monks area of the present district is similar to that around Ingham, demonstrating that no significant dilution of quartzite had occurred below the supposed confluence with the Kesgrave river. It is likely, therefore, that the quartzite-rich gravels belonged to a stream which remained separate from the Kesgrave river (Hopson and Bridge, 1987).

The Bytham Sands and Gravels are restricted to the north-west of the district, where they roughly follow the course of the Waveney valley, being recorded in boreholes from Swan Green [29 75] to Hedenham [31 94] and near Toft Monks [43 93]. Gravels are recorded with base levels between 2.2 and 30.5 m above OD, and upper surface levels between 6.3 and 32.5 m above OD. They can be divided into three terraces (Figure 19a), as in the Diss district to the west (Mathers et al., 1993, fig. 5), and these appear to correspond to the Ingham, Knettishall and Timworth gravel members of Lewis (1994): his highest terrace, the Seven Hills Gravel Member, does not extend eastward into the district. The distribution of the three terraces in the present district indicates that they were formed in a valley surprisingly similar to the present Waveney valley, and gives no support to the traditional theory (Whiteman, 1992, fig. 2; Arthurton et al., 1994, fig. 29) that they formed as a part of the Kesgrave Group in a much broader valley orientated south to north. South of Swan Green, borehole records on sheet TM 27 SE show Crag at levels as high as 35 m above OD, and it appears likely that these are beyond the southern limit of the Bytham Sands and Gravels. East of Bungay no Bytham Sands and Gravels are known higher than 13.1 m above OD, but eastwards from Barsham Hill [40 89] Corton Formation sands have been mapped resting upon Crag at heights around 15 to 20 m above OD, so clearly there the limit of the Bytham Sands and Gravels lies beneath the present floodplain of the Waveney; east of Toft Monks the gravels are absent but boreholes record the Cromer Forest-bed Formation, which is believed to be laterally equivalent to the latest Bytham Sands and Gravels. Quartz-rich gravels recorded in field brash around Cookley [34 75] may represent the south-western limit of the deposits.

As in the case of the Kesgrave Group, the Bytham Sands and Gravels have not been delineated on the maps because of the difficulty of distinguishing them from the overlying glacial sediments in the field. In practice, in the Waveney valley they are included within the Beccles Beds on the 1:10 000 scale maps and within the Corton Formation on the 1:50 000 map, whereas around Cookley [34 75] they are included within the outcrop of sands and gravels of the Lowestoft Till Formation.

Stratigraphy and sedimentation

The deposits were probably laid down during successive cold climatic periods, by a major braided river that flowed from the West Midlands and southern Pennines across the area now occupied by Fenland, crossed the Chalk outcrop at Bury St Edmunds, and thence followed a course roughly along the present Waveney valley. Rose (1987) showed the base of the deposits falling from around 90 m above OD near Stratford upon Avon to near OD at Diss. In the present district quartz-rich gravels are recorded over a height range of 2.2 to 32.5 m above OD, which falls within the altitude limits quoted

for the Bytham Sands and Gravels in the Diss area (Rose, 1987).

The deposits largely comprise pebbly sands and sandy gravels, commonly clayey. The gravel fractions vary through fine to coarse and include rounded and angular flint, rounded vein quartz, and both grey and purple quartzite, with traces of chalk, silicified limestone, shells, igneous and metamorphic rocks, mica, iron pan and ironstone. The sand fractions may be fine to coarse and are dominated by angular to rounded quartz, with a little quartzite and angular flint. They are most commonly yellowish brown or yellowish orange, although shades of olive and grey are also common. Thin laminated seams of clay and silt, in various shades of yellow, orange, brown or olive, are scattered through the deposits.

Along the valley of the Waveney from Bury St Edmunds to Diss (Clarke and Auton, 1982; Mathers et al., 1993) two distinct deposits have in the past been recognised, one lithologically equated with the Kesgrave Group, the other, the Ingham Sand and Gravel, being more quartzite-rich. The Ingham Sand and Gravel was shown by Mathers et al. (1993, fig. 5) as the middle one of three terraces, the highest terrace being termed Kesgrave Sands and Gravels. However, Hopson and Bridge (1987, fig. 4) demonstrated that even the so-called Kesgrave Sands and Gravels here can be distinguished on flint: quartz:quartzite ratios from the true Kesgrave Group to the south. The gravel fraction of the Ingham Sand and Gravel is characterised by quartz and quartzite derived from the Trias of the West Midlands, in particular by liver-coloured 'Bunter' quartzites derived from the Kidderminster Formation. Chert occurs in small amounts, most probably from the Carboniferous of the south Pennines (Rose, 1987). The Ingham Sand and Gravel sensu stricto also has a noticeably low content of rounded flints, indicating a lack of input from the Palaeogene rocks of the London Basin.

Table 6 shows all pebble-counts for gravels interpreted as post-Crag and pre-Anglian in an industrial minerals survey of a wide area that extended into the Waveney valley of the present district upstream of Mettingham (Auton et al., 1985). Also shown is a plot of the average of results obtained from later boreholes in the Toft Monks area of the present district (Hopson and Bridge, 1987). Plotting the ratios of flint, quartz and quartzite for these deposits (Figure 19c) shows a single population with widely varying ratios, and does not distinguish 'Kesgrave' and 'Ingham' lithologies on different terraces, as was claimed by Mathers et al. (1993) for the area upstream.

Within the present district, quartzite:quartz ratios in the Bytham Sands and Gravels vary relatively evenly

Figure 19 Bytham Sands and Gravels in the district.

a. Boreholes proving pre-Anglian quartz/quartzite-rich gravels indicating three terrace levels within the Bytham Sands and Gravels. The Cromer Forest-bed Formation is also included.

b. Height ranges of Bytham and Cromer Forest-bed deposits in boreholes depicted in a, indicating three terrace levels.

c. Ratio plot of pebble counts for quartz, quartzite and flint for +8-16 mm fractions.

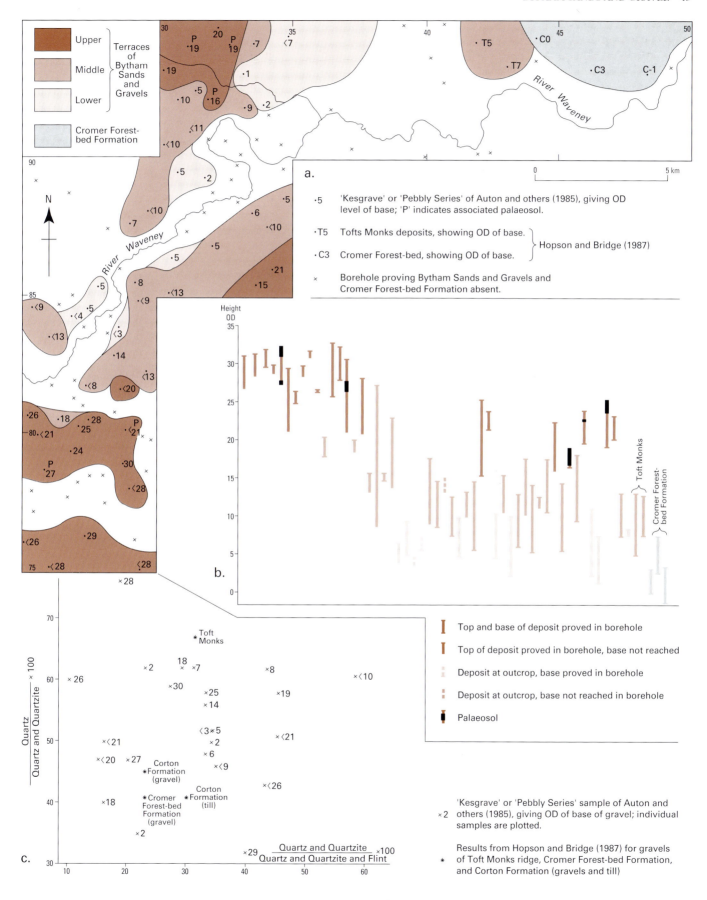

a.

Terraces of Bytham Sands and Gravels
- Upper
- Middle
- Lower

Cromer Forest-bed Formation

·5 'Kesgrave' or 'Pebbly Series' of Auton and others (1985), giving OD level of base; 'P' indicates associated palaeosol.

·T5 Tofts Monks deposits, showing OD of base.

·C3 Cromer Forest-bed, showing OD of base.

⎫ Hopson and Bridge (1987)

× Borehole proving Bytham Sands and Gravels and Cromer Forest-bed Formation absent.

b.

I Top and base of deposit proved in borehole

I Top of deposit proved in borehole, base not reached

⦙ Deposit at outcrop, base proved in borehole

⦙ Deposit at outcrop, base not reached in borehole

▮ Palaeosol

c.

×2 'Kesgrave' or 'Pebbly Series' sample of Auton and others (1985), giving OD of base of gravel; individual samples are plotted.

* Results from Hopson and Bridge (1987) for gravels of Toft Monks ridge, Cromer Forest-bed Formation, and Corton Formation (gravels and till)

between 0.5 and 3.2, with 2.0 for the Toft Monks samples, while Ingham Sand and Gravel samples quoted by Auton et al. (1985) and by Clarke and Auton (1982) vary between 0.7 and 11.5, giving a considerable overlap with samples from the present district. However, analysing the samples on the basis of separate terraces of the Bytham river gives average results for the present district of 1.0, 1.3 and 1.3 for the lower, middle and upper terraces, while the quartzite:quartz ratios for the four terraces distinguished by Lewis (1993) are similarly almost identical one to another. These results suggest that in reality the relative supplies of quartz and quartzite remained constant throughout the history of the Bytham river, and must cast doubt on the existence of the Ingham Sand and Gravel as a separate deposit: the term may merely refer to the quartzite-rich individuals within a variable population of samples.

The ratio of flint to quartz plus quartzite varies in the present district from 9.0 to 0.7; there is considerably more flint in the samples from the present district than in those from farther upstream, but this might be expected since the Bytham river would have picked up flints as it flowed over a surface of Crag containing flint gravels: Auton et al. (1985) record several instances of flint gravels (Westleton Beds) lying at the top of the Crag. Rounded flints, such as might be derived from the Crag, are present in the samples of Bytham Sands and Gravels from the present district in amounts varying between 7 and 41 per cent, compared to between 2 and 11 per cent for the Ingham samples of Auton et al. (1985).

Examination of the fine sand mineralogy of the Bytham Sands and Gravels from sites in Warwickshire, Lincolnshire, Norfolk and Suffolk (but not from the present district) by Bateman and Rose (1994) demonstrated derivation primarily from the Triassic sandstones of the eastern West Midlands, with minor contributions from the Carboniferous rocks of the southern Pennines and from Jurassic rocks.

Auton et al. (1985) recorded clay-enriched palaeosols developed in the highest terrace recognised in the Bytham Sands and Gravels of the district. These comprise sandy silts, sandy clays and clayey pebbly sands with pebbles of flint, quartz, quartzite and red-stained sandstone, and may be salmon-pink, reddish brown, orange or yellow.

The wide range of altitude over which quartzose gravels are found in the upper part of the Waveney valley implies that deposition occurred over a long period of time, as in the case of the Kesgrave Group, and it is likely that the Bytham and Kesgrave deposits span roughly the same time ranges. This is in any case likely since the deposits were formed in similar braided river complexes, albeit in separate drainage systems, and the presence of palaeosols in the Bytham Sands and Gravels also supports correlation with the Kesgrave Group. However, it would be unreasonable to attempt an exact correlation of terraces between the two groups on the evidence currently available.

An isolated borehole near Toft Monks [428 948] (Hopson and Bridge, 1987) shows the Bytham Sands and Gravels overlain by the Rootlet Bed of the Cromer Forest-bed Formation. Farther east this formation, including the Rootlet Bed, is recorded within a depression (the Corton Embayment of Hopson and Bridge, 1987) bevelled into the ridge on which the gravels of Toft Monks rest. Three boreholes within the Cromer Forest-bed of the Corton Embayment are shown on Figure 19, and demonstrate that it lies at the level of the lowest of the three terraces within the Bytham Sands and Gravels. The implied correlation suggests that this lowest terrace is likely to be Cromerian in age. The Toft Monks samples (Figure 19) align with the middle terrace, and hence apparently with the 'Ingham' terrace of the Diss district (Mathers et al., 1993) and the Knettishall Gravel Member of Lewis (1993). All are likely to be pre-Cromerian and may thus be Beestonian in age (Hopson and Bridge, 1987).

The highest terrace in the district, which appears to correlate with the high-level 'Kesgrave Sands and Gravels' of Mathers et al. (1993) in the Diss area and with the Ingham Gravel Member of Lewis (1994), is considerably higher than the lower two terraces and could thus be significantly older. This highest terrace is the only one known to be associated with palaeosols (Figure 19), and these accord with a long and complex history for the terrace. In two cases gravels occur both below and above a palaeosol, implying two periglacial periods of gravel accumulation separated by a warm period, whilst in one case (Figure 19c; this site lies in the Diss district) two palaeosols are recorded, below and above a gravel, implying two warm periods.

The Seven Hills Gravel Member of Lewis (1993), which does not extend into the present district, is the highest (and hence oldest) terrace known along the whole length of the Bytham river. Lewis (1993) suggested that it could be as old as Pre-Pastonian a/Baventian, but this seems doubtful since the Westleton Beds of the Crag Group, which are now believed to be of that age, contain very little quartz and quartzite; the Seven Hills Gravel Member may alternatively date from the succeeding cold period. It is believed that at no time did the proto-Thames and the Bytham river join within the present district, and it follows that during those periods when sea-levels were sufficiently high for marine deposition to occur within the district, for instance during the Cromerian, the rivers had separate mouths. However, when sea-levels were low the rivers would have united to the east of the district.

Details

Waveney valley upstream from Bungay

Figure 19a shows the sites of all boreholes drilled in the district during a survey of the Waveney valley (Auton et al., 1985) and which were interpreted by those authors as including post-Crag pre-Anglian gravels (their 'Kesgrave Sands and Gravels' and 'Pebbly Series'). The map shows the OD level of the base of the deposit in each borehole. Where such deposits were proved to be absent a cross (x) is shown, and where palaeosols were proved these are indicated by a letter P. The height ranges of the deposits at each borehole are shown in Figure 19b, while Table 6 details all pebble counts made for the deposits. Auton et al. (1985) did not record any Ingham Sand and Gravel within the district, although some of their pebble counts give quartz:quartzite ratios comparable to samples which they described as Ingham Sand and Gravel farther west. Figure 19c is a ratio plot demonstrating the relative flint, quartz and quartzite contents: again the OD levels are shown. The results show a wide range of relationships, and a wide range of OD levels, but no significant correlations emerge. Plotting these same factors against angular flint percentages also gives no obvious correlation.

Auton et al. (1985) distinguished the 'Pebbly Series' from the 'Kesgraves' on the basis of the former having a coarser gravel fraction, more angular flints and a lower proportion of quartz and quartzite. For this reason individual pebble counts are quoted in Table 6 instead of averages. However, the data presented show no distinction between 'Pebbly Series' and 'Kesgraves', either in terms of percentages of angular flint or of quartz and quartzite, or in terms of coarseness of the deposit: mean gradings for all samples of 'Kesgraves' are 8% fines, 72% sand and 20% gravel, compared to 7%, 69% and 24% for the 'Pebbly Series'. On the contrary, the data imply a complex of Bytham terraces, probably with much reworking from one terrace level to another. There remains the possibility that some of the deposits considered here are in reality Corton Formation (Anglian), but the fact that the deposits rest upon a series of terraces falling north-eastward within the Waveney valley argues against this, since Corton Formation gravels would have been derived from the north.

Within the present district and west of easting 35, 38 boreholes drilled for sand and gravel assessment record 'Kesgrave Sands and Gravels' (up to 9.2 m thick) or 'Pebbly Series' (up to 9.9 m thick) or both. Both units would now be included in the Bytham Sands and Gravels. Four boreholes within the highest terrace record palaeosols, with base levels between 17.4 and 26.4 m above OD. In two of them (TM28SE/44 [2884 8017] and TM39SW/36 [3105 9407]) the palaeosol separates two beds of gravel, implying a complex origin for the terrace, with two periods of periglacial terrace aggradation separated by a warm phase. In the other two (TM39SW/43 [3261 9411] and TM39SW/44 [3183 9205]) a single palaeosol overlies gravel.

The deposits have not been distinguished on the published maps, but they may occur within the areas mapped as Corton Formation in the Waveney valley, since at Hulk's Grave [2629 8026] near Withersdale Street, just to the west of the district, Lawson (1982) recorded a soil sample with a gravel content of 43.3% rounded flint, 25.9% angular flint (mainly fractured rounded clasts), 15.7% quartz and 7.4% quartzite. The low proportion of unrounded flints would appear to exclude a glacial origin, while the content of quartz and quartzite is rather high for Crag. This sample was collected at about 30 m above OD.

In the flooded pit at Fox Burrows [2945 9365] up to 3.5 m of pale orange and yellowish brown, medium-grained, well-sorted, slightly micaceous, laminated and ripple-drift bedded sands include, near the top of the section, some well-rounded gravel with up to 63% of quartz and quartzite pebbles. At an old brickyard at Hedenham, trial pits and an auger borehole [3045 9310, 3022 9317] revealed glacial gravel overlying up to 1.2 m of rounded gravels with abundant quartz and quartzite pebbles, resting on Crag. An old pit [3044 9362] 500 m to the north showed glacial sediments overlying 6.5 m of sands which contained a gravel with an abundance of rounded quartz and quartzite (50%) and subrounded flint (37%). In contrast, rounded-flint gravels underlying glacial gravels in a nearby ditch section [3054 9356] comprised 93.2% flint and only 5.9% quartz and quartzite, and hence are probably Crag.

Bungay to Beccles

Several boreholes and exposures record sands and gravels beneath Corton Formation (Anglian) till (Wilcox and Horton, 1982), but there is no way of determining whether they refer to the Bytham Sands and Gravels or to the basal Anglian deposits. A pit at Leet Hill [3807 9294] revealed 3.2 m of cross-bedded sandy gravel, rich in quartz and quartzite, with thin lenses of medium- to coarse-grained sand. This was overlain by gravels of the Corton Formation, characterised by erratics of Scandinavian origin (Hopson, 1991). An auger hole within this pit (Wilcox and Horton, 1982) indicated an additional 5.5 m of sandy gravel overlying Crag, implying a thickness of 8.7 m of Bytham Sands and Gravels at this site.

Beccles to Burgh St Peter

In 1983, some thirty boreholes were drilled at about 2 km intervals between Beccles and Corton, mainly to the north of the present district (Hopson and Bridge, 1987). Of these, five within the district proved either Bytham Sands and Gravels (boreholes TM39SW/135, 136) or Cromer Forest-bed Formation (TM39SW/137, TM39SE/8, 9). The upper surface of the Crag was found to form a low ridge rising to 9.1 m above OD in the Toft Monks area, a broad embayment (the Corton Embayment) at about OD around Burgh St Peter, and a north-easterly trending linear depression (the Raveningham–Lound Depression) descending to 9.3 m below OD in the adjacent district to the north.

On the Toft Monks Ridge, 'Bunter'-rich quartzose gravels (Bytham Sands and Gravels) up to 7 m thick rest directly on Crag at between 5 and 8 m above OD. The deposit varies from pebbly sand to gravel, and pale quartzite pebbles apparently derived from the 'Bunter' of the Midlands are the most common constituent after flint, comprising some 20% by weight of the gravel fraction (Table 6). In the Corton Embayment, two units of sand and gravel are separated by the Rootlet Bed of Blake (1884, 1890), so are included here in the Cromer Forest-bed Formation. Pebble composition data (Table 6, Figure 19) reproduced from Hopson and Bridge (1987, table 2) demonstrate that the ratio of quartzite to quartz in the Toft Monk gravels is comparable with those of Ingham: the higher proportion of flint in the Toft Monk gravels may be accounted for by downstream dilution between Ingham and the present district.

Cookley

Quartz and quartzite pebbles are common in the field brash around Cookley [34 75] at levels of 22 to 26 m above OD, and may indicate the presence of Bytham Sands and Gravels beneath sands and gravels of the Lowestoft Till Formation. They have not been noted farther east along the valley towards Halesworth, so this may be the farthest south-east that the deposit occurs.

'BECCLES BEDS'

The 'Beccles Beds' (Wilcox and Horton, 1982; Horton, 1982a, b; Auton et al., 1985), or 'Beccles Formation' (Wyatt, 1981), was an informal name used in the Waveney valley for all the deposits between the top of the Norwich Crag and the base of the Lowestoft Till Formation. The unit thus included the Bytham Sands and Gravels and the Corton Formation, which in that area included the Starston Till (Lawson, 1982) as well as sands and pebbly sands. The base of the unit was taken at the first appearance of large numbers of pebbles in the sequence, and the top was defined at the appearance of deposits rich in chalk clasts (the Lowestoft Till Formation). The term proved useful for mapping these poorly exposed sediments and for the classification of existing records of inadequately described boreholes and wells, at a time when the extent and relationships of the Bytham Sands and Gravels and the Corton Formation were not understood. Its use has now been discontinued, though it survives in some publications. Thus, within the district, Beccles Beds are shown at outcrop on 1:10 000 scale maps TM 38 NE, NW, 39 SE and SW, and appear in borehole logs in the margin of TM 38 SW, although they are not present at outcrop on that sheet. Sheets TM 28 NE, SE and TM 29 SE were surveyed before the term Beccles Beds was introduced: on the 1:10 000 scale maps the outcrop is not separated from the sands and gravels of the Lowestoft Till Formation, nor is the name Beccles Beds used, but the Beccles Beds are discussed in the accompanying reports (Wyatt, 1981; Lawson, 1982).

FOUR

Quaternary (Anglian to Devensian): glacial and interglacial deposits

Deposits of Anglian age form the higher ground of the district, with older formations at outcrop in river valleys; in low-lying eastern areas, and offshore, the Anglian deposits have been removed by erosion and younger deposits rest directly on earlier Pleistocene strata. All of the Anglian deposits are ultimately of glacigenic derivation, and appear to have been formed during a single glacial period: evidence for the onset of cold climatic conditions is preserved in the Cromer Forest-bed Formation underlying the Anglian deposits at Corton, in the Great Yarmouth district, where Blake (1884b) noted vertical to subvertical wedge shaped structures which were later recognised as a polygonal pattern of ice-wedge casts developed in early Anglian times (Gardner and West, 1975).

The Anglian glacial deposits are included within two formally defined units (Arthurton et al., 1994), the Corton Formation below and the Lowestoft Till Formation above. These derive from two separate ice-sheets: the 'North Sea Drift' (Harmer, 1909, 1910a, b) or 'Scandinavian Ice Sheet', which entered the area from the north or just east of north (Perrin et al., 1979) and was responsible for the deposits of the Corton Formation, and the 'British Eastern Ice Sheet', which entered from the west (Perrin et al., 1979; Hart et al., 1990) and was responsible for the Lowestoft Till Formation. The deposits of the Corton Formation are characterised by a suite of Scandinavian igneous and metamorphic erratics, while those of the Lowestoft Till Formation are derived from Mesozoic outcrops to the north-west, principally the Chalk and Kimmeridge Clay.

Cox and Nickless (1972) argued that the two Anglian ice sheets coexisted in the Norwich area, one depositing Lowestoft till in the west, and the other depositing Corton till in the east, with an outwash zone of sand and gravel between them. They suggested that the British Eastern Ice Sheet continued south-eastwards to override the now-stagnant Scandinavian ice. Perrin et al. (1979) also considered the two ice-sheets to have coexisted, with at least a part of the Marly Drift of north-west Norfolk being a product of a complex interaction between them. Hart and Peglar (1990), working on glacial landforms in north-east Norfolk, and Hart and Boulton (1991), also considered the ice-sheets to have coexisted. Lewis (1993) found evidence that the ice-sheets coexisted at Knettishall and Barnham, near Thetford. However, Hopson and Bridge (1987) considered that the degree of dissection of the Corton Formation suggested that it had been affected by erosion for a considerable period before being overriden by the British Eastern Ice Sheet: the eroded top of the sands of the Corton Formation falls from about 25 m above OD near Norwich to 10 m at Corton, and the Lowestoft Till rests on successively lower levels of the sands, locally coming to rest on the Crag at 1.2 m below OD. Mathers et al. (1993) supported

this conclusion but Wilcox and Horton (1982) recorded a transitional junction between the Corton and Lowestoft till formations.

When the Anglian ice melted, the present-day drainage pattern developed on the predominantly till-covered landscape. Some of the rivers reoccupied their preglacial channels, while others utilised the new topography of meltwater channels. For example, the River Waveney appears to have broadly followed its preglacial north-easterly course in the west of the district, but switched to a more easterly aligned marginal meltwater channel farther downstream.

During cold periods sands and gravels were deposited as river terrace deposits within the Waveney valley, and also on a more limited scale along the Blyth, Deben and Alde valleys. Evidence suggests that these sands and gravels may range in age from late Anglian to late Devensian.

During the intervening warmer periods (Hoxnian and Ipswichian), interglacial silts and clays were deposited within hollows on the land surface. Several such sites have been identified within and adjacent to the present district.

Head, locally gravelly, mantles many of the lower slopes, and floors minor valleys. It accumulated mainly by solifluction in a periglacial environment during Devensian times, but smaller accumulations have continued to form, by the processes of hill-wash and soil-creep, up to the present day.

The cover silt, a veneer of silt or fine-grained sand which covers much of the adjacent ground to the north, is absent or very thin within the present district, and has not been mapped as a separate deposit. In the Great Yarmouth district it overlies all the Quaternary deposits apart for the Holocene Breydon Formation, and is therefore assumed to be Devensian in age.

Within the Waveney valley in the adjacent Great Yarmouth district, a suite of gravels, known as the Yare Valley Formation, has been recorded extensively beneath the Holocene alluvial deposits. These gravels are known to continue into the present district but because of a lack of borehole data their full extent is unknown. The age of these gravels is uncertain; it is probable that more than one suite is present. Some may be late Anglian and derived from glacial outwash; others may be Devensian and represent the buried component of the first (Floodplain) terrace.

CORTON FORMATION

Nomenclature and distribution

The Corton Formation is named after a coast section at Corton [5451 9722] (Figure 20) in the adjacent Great Yarmouth district, where it forms the lower part of the

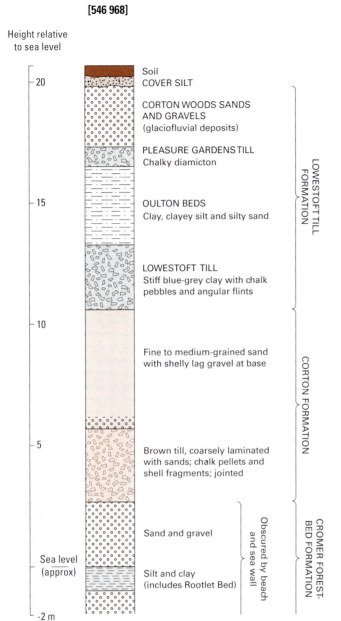

Figure 20   The generalised succession at Corton.

type section for the Anglian Stage (Banham, 1971; Mitchell et al., 1973). The upper part of this section is in the Lowestoft Till Formation and the lower part is the stratotype for the Corton Formation (Arthurton et al., 1994). The Corton Formation is synonymous with the North Sea Drift Formation of Mathers et al. (1987) and the North Sea Drift Group of Perrin et al. (1979) and Hopson and Bridge (1987). It is equivalent to the Contorted Drift (Wood and Harmer, 1868) of north-east Norfolk. The deposits of the Corton Formation owe their origin to the Scandinavian Ice Sheet, in the first major ice advance known to affect this area. It entered East Anglia from the north and north-east, having travelled across what is now the North Sea.

The deposits are characterised by Scandinavian erratics and material from the North Sea Basin. The formation is a complex unit which in the present district comprises sands with subordinate lenticular tills and pebbly sands. In this memoir, informal facies terms are used for the component lithological units. However, all have in the past received formal names (Table 7). The till-dominated parts of the formation have been referred to as Lower Boulder Clay (Gunn, 1867), Lower Glacial (Wood and Harmer, 1868), Norwich Brickearth (Harmer, 1902; Boswell, 1916; Hopson, 1991), or North Sea Drift (Harmer, 1910; Perrin et al., 1979); and as Cromer Till (Baden-Powell, 1950; Banham, 1971) or First Cromer Till (Banham, 1970) in the north-east of the area. Boswell (1916) used the term Cromer Till for the deposit when fresh, and Norwich Brickearth when weathered and decalcified. However, till in the north-west of the district, which is greyer and contains more chalk than that at Corton, has been referred to as Starston Till (Lawson, 1982), with a type section [2427 8444] at Starston where the till is 3.5 m thick. It is not certain that the Starston Till is the direct equivalent of the till seen in the type section at Corton, though this appears likely.

The sands (Plate 8) which make up the bulk of the formation have been termed (Table 7) Middle Glacial (Wood and Harmer, 1868), Corton Beds (Baden-Powell, 1948, 1950; Banham, 1971) and Corton Sands (Baden-Powell, 1950; Bridge and Hopson, 1985; Hopson and Bridge, 1987). In the west of the district these same sands have been named Mendham Beds (Auton et al., 1985), with a type section at Mendham Pit [2716 8245]. Pebbly beds within the sands have been named Leet Hill Sands and Gravels (Bridge and Hopson, 1985; Hopson and Bridge, 1987).

Wood and Harmer (1877) considered that the 'Contorted Drift' (North Sea Drift) extends south into Suffolk, but Boswell (1914) concluded that its southern limit lies close to Beccles and Lowestoft. This is confirmed by the present survey, which has found the Corton Formation to occur only in the northern part of the district: Figure 18 indicates the approximate southern limit to which the deposits have been proved. This line may not represent the original extent of the Corton Formation since considerable erosion is likely to have occurred before the deposition of the overlying Lowestoft Till Formation. The deposits overlie the Bytham Sands and Gravels in the north-west and the Cromer Forest-bed Formation in the north-east (Figure 18), overstepping southwards on to the Crag.

Despite the widespread distribution of the Corton Formation, and its considerable thickness (at least 20 m in the north of the district), the outcrop is small because of the masking effect of the Lowestoft Till Formation. Apart from the lowland areas between Lowestoft and Pakefield, and around Wrentham and South Cove, the formation crops out only as narrow strips along valley sides and in the sea cliffs. It is best exposed in the coastal sections, particularly from Lowestoft to Kessingland and south of Covehithe, where all its lithologies can be examined. The coast from Kessingland northwards was

Table 7 Lithostratigraphical nomenclature adopted for the Anglian glacial deposits of the district, compared with a selection of earlier schemes.

THIS SURVEY	ARTHURTON ET AL. (1994)	HOPSON AND BRIDGE (1987)	BANHAM (1971)	BADEN-POWELL (1948; 1950)	WOODWARD (1881); BLAKE (1880)	WOOD AND HARMER (1868); WOOD (1880)
Lowestoft Till Formation includes Pleasure Gardens Till, Oulton Beds, Corton Woods Sands and Gravels, Haddiscoe Sands and Gravels, Aldeby Sands and Gravels, and un-named glaciofluvial sands and gravels	Glaciofluvial Deposits includes Haddiscoe Sands and Gravels and Corton Woods Sands and Gravels	Plateau Gravels Valley flank sands and gravels			Gravel	
	Lowestoft Till Formation includes Pleasure Gardens Till and Oulton Beds and un-named sands and gravels	Pleasure Gardens Till Oulton Beds Lowestoft Till	Pleasure Gardens Till Oulton Beds Lowestoft Till	Lowestoft Boulder Clay	Boulder Clay = Chalky Boulder Clay	Upper Glacial = true Boulder Clay = Chalky Boulder Clay
Corton Formation including till layers	Corton Formation including till layers	Corton Sands Leet Hill Sands and Gravels	Corton Beds	Corton Beds = Corton Sands		Middle Glacial
		North Sea Drift = Norwich Brickearth	First Cromer Till	Cromer Till	Stony Loam = Brickearth = Norwich Brickearth	Lower Glacial

much better exposed a century ago when active cliff erosion maintained a fresh cliff face; the section was measured in detail by Blake (1884a, b). Inland the formation is not commonly exposed except in degraded sand and gravel pits, but it is well known along the Waveney valley from two sets of BGS boreholes. One set was drilled for sand and gravel assessment of an area extending westward from Bungay (Auton et al., 1985), and the other was drilled between Corton and Beccles as part of a regional survey of East Anglia (Bridge and Hopson, 1985; Hopson and Bridge, 1987).

Throughout the area in which the Corton Formation is present, sands are the dominant lithology, ranging up to 15 m thick, and on the published map the formation is shown undivided except where layers of till have been mapped out. The pebbly facies of the formation ('Leet Hill Sands and Gravels') is not developed at the type locality at Corton but is widespread inland in the Great Yarmouth district, reaching a thickness of 10 m in a linear depression cut into the Crag surface (Hopson and Bridge, 1987) and present both below and above the till sheet (Bridge and Hopson, 1985). It is less well developed in the present district, except around Leet Hill, the type locality for the 'Leet Hill Sands and Gravels' (Hopson and Bridge, 1987). It has nowhere proved possible to map the pebbly sands as a separate unit.

In the Great Yarmouth district to the north, Corton till forms a single continuous sheet resting on an inclined surface which rises gently inland from 0 to 5 m above OD at Corton to over 13 m above OD near Toft Monks (Hopson and Bridge, 1987). Mapping and borehole data show the sheet to be impersistent locally in the southern part of the district (Arthurton et al., 1994). This till has been tentatively correlated with the lowest of three Cromer Tills in the Norfolk succession (Bridge and Hopson, 1985). In the Norwich district a single till sheet ('Norwich Brickearth'), ranging in thickness from 3 to 6 m, represents Corton till. In the present district only sporadic occurrences of till have been proved, up to at least 7.1 m thick in boreholes in the north-west of the district, and there does not seem to be a continuous sheet. In three boreholes in the Waveney valley (TM 27 NE/19, TM 38 NW/31, TM 39 SW/52 [2808 7752, 3070 8948, 3478 9445]) two tills have been tentatively identified within the Corton Formation; it appears likely that

Plate 8 Cross-bedded sands of the Corton Formation overlain by Lowestoft till [5366 8857] (GS 567). Height of exposure about 5 m.

all such occurrences within the present district can be correlated with the single till sheet of the Great Yarmouth district.

Stratigraphy and sedimentology

The Corton Formation is dominated by sands, which are well exposed along the coast, particularly from Kessingland to Lowestoft, and are well known from boreholes along the Waveney valley. Till and pebbly sands are restricted to the lower part of the formation and are not widely exposed, although both are well known from the Waveney valley boreholes. In the north-east of the district, Hopson and Bridge (1987) found that the pebbly sands were thin and absent where the till was present, but several boreholes in the north-west of the district record both facies, and both are recorded in the coast section (Blake, 1884a). The till, where present, does not always lie at the base of the succession as it does in the type section at Corton, but may be underlain by gravel and/or sand. Gravels may occur above, below, or above and below the till.

Corton sand

The Corton sands are well sorted, fine to medium grained, locally clayey, formed from subangular to sub-rounded quartz with subsidiary sand-grade flint, quartzite and disseminated chalk grains, calcite prisms and some mica flakes. Abraded shell debris occurs in the coarser grades. The sands are most commonly greyish orange or yellowish brown ('buff'), but olive-grey and other shades of yellow, orange and brown are recorded. Thin layers or laminae of silt, clay or pebbly diamicton occur, but

pebble-grade material accounts for only 0.3% of the deposits (Hopson and Bridge, 1987), occurring as stringers of gravel, generally fine grained, with angular and rounded flint, vein quartz, quartzite, chalk, and traces of Scandinavian erratics including porphyry, granitoids and metamorphic rocks. Analyses of these pebbles in the Beccles and Burgh St Peters area indicated 64% angular and subangular flint, 20% quartz and 5 to 10% each of rounded flint and quartzite (Hopson, 1991). An analysis of the 8 to 16 mm fraction of a single sand sample from the west of the district (in borehole TM38 NW/43 [3351 8803]) indicated 51% angular flint, 11% rounded flint, 20% vein quartz and 18% quartzite (Auton et al., 1985). The presence of Scandinavian erratics in significant quantities demonstrates that these sands are a part of the Corton Formation and not of the overlying Lowestoft Till Formation (Hopson and Bridge, 1987).

Low-angle to horizontal lamination and ripple-drift lamination are recorded (Wilcox and Horton, 1982), while cross-bedding structures with cross-sets up to 1.5 m high are well displayed in the coast sections (Bridge and Hopson, 1985). Both large- and small-scale sedimentary structures indicate that the sands are water-laid, with flow dominantly from the north and north-west, although the high degree of sorting and roundness of the sand grains implies that at least a proportion of the material has also had an aeolian history. The sands are locally micaceous, for instance at the old pit at Watch House Hill [356 900] (Wilcox and Horton, 1982). They can be distinguished from Crag sands in the field by their more uniform pale yellow colour, the local presence of chalk, the lack of iron 'pan', and the smoother 'feel' of the grains that results from their higher degree of rounding.

On the basis of boreholes in and near the north-east of the district, Hopson and Bridge (1987) split the sands of the Corton Formation into three fining-upward sedimentary cycles culminating in silts and clays. These cannot be recognised from field mapping (Hopson, 1991), and were not detected during the survey of the present district. The cycles were considered to represent successive stages in the infilling of a pre-existing topography. Only the lowest cycle was recorded at Corton, hence the stratotype does not contain the full succession. This first cycle was found in boreholes south-east of Toft Monks [43 94] and infills a basinal low in the Corton Embayment, but does not overtop the Bytham Sands and Gravels of the Toft Monks ridge; its top was placed at a planar disconformity and zone of cryoturbation at around 10 m above OD. Sands of the second cycle are more extensive and were recognised at Leet Hill pit [3807 9294]; the top of this cycle is marked by clays and disturbed bedding at around 20 m above OD. The third cycle, finer in grade than the lower two, was recognised in boreholes TM 49 SW/134 [4009 9337] and TM 49 SW/135 [4179 9423] west of Toft Monks and in overgrown pits around Gillingham Thicks [421 935] and Upland Farm [403 935]. The cycles were correlated with those represented in north-east Norfolk by the First, Second and Third Cromer Tills of Banham (1971).

The junction between the sands and the overlying waterlaid basal bed of the Lowestoft Till is regarded as transitional by Wilcox and Horton (1982), who noted temporary sections along the Beccles Bypass in which the sands pass up into till with large dropstones. In contrast, Bridge and Hopson (1985) recorded sporadic calcareous cementation (calcrete) up to 30 cm thick in the top 2 m of the sands, and derived fragments of the calcrete in the lowest metre or so of the Lowestoft Till. By implication, the calcrete was formed in the Corton Formation before this was eroded by the ice sheet that deposited the Lowestoft till. Woodward (1881) explained the formation of the calcretes by percolation of calcium carbonate from the overlying till, but this could not be the case if the calcretes were formed before the till. The conflict of evidence has not been explained.

The presence of chalk grains in the sand is difficult to account for, since the underlying till, at least in the east of the area, is low in chalk or chalk-free. Hopson and Bridge (1987) suggested that the explanation might be that the North Sea ice incorporated much Palaeogene material without deeply eroding the Chalk, while the outwash streams were actively eroding Chalk bedrock. However, the sands of the Corton Formation are not considered to be fluvial (see below), and a more likely explanation is that the sand-grade chalk was derived from the British Eastern Ice Sheet, which most workers agree coexisted with the North Sea ice.

Two samples (MPA 40 015, MPA 40 016) from the Corton sands at Leet Hill [378 929] were examined for dinoflagellate cysts. Both samples proved barren of cysts, but contained quantities of plant tissue including cuticular remains and tracheids together with fungal hyphae and spores (Harland, 1993).

Corton pebbly sand

The sand and gravel survey of the Waveney valley revealed very few true gravels within the Corton Formation, though pebbly sands are common. The sand fraction is generally fine to medium grained and comprises angular to subrounded grains of quartz and flint with a trace of chalk. The colour is most commonly yellowish brown, but yellowish orange, pale brown, greyish orange and olive-grey are recorded. The gravel fraction is fine to coarse grained and dominated by flint pebbles which vary from rounded through subrounded and subangular to angular. Vein quartz and quartzite are common, with chalk as a minor constituent in most samples. Iron pan, ironstone, shell fragments and limestones are rare. Silt beds and clay partings are recorded.

Two samples of true gravels from boreholes in the west of the district (TM 28 SE/40, 41 [2857 8391, 2838 8287]) comprised 54 to 59% angular flint, 13 to 16% rounded flint, 6 to 9% vein quartz, 14 to 15% quartzite, up to 2% of chalk and other limestones, and up to 5% of other species (Auton et al., 1985). Gravels from boreholes in the north-east of the district (Bridge and Hopson, 1985; Hopson and Bridge, 1987) showed the proportions of flint, quartz and quartzite to be within the range obtained for the Bytham Sands and Gravels, but to include rather more quartzite than the Corton till or the Cromer Forest-bed (Table 6; Figure 19c). Bridge and Hopson consider that the Corton Formation gravels contain a large amount of material derived from the Bytham Sands and Gravels, but can be distinguished from them by their content of Scandinavian granites, purple rhomb porphyries and metamorphic rocks. This suggested derivation was supported by the heavy mineralogy of the gravels (Bridge and Hopson, 1985), which contain high levels of zircon and rutile like the pre-Anglian gravels. At a quarry at Leet Hill [3807 9294] gravels of the Corton Formation (the 'Leet Hill Sands and Gravels' of Hopson and Bridge, 1987) closely resembled the underlying Bytham Sands and Gravels in overall appearance, but the 4 to 8 mm fraction was found to contain purple rhomb porphyries, with granitic and metamorphic rocks, all of Scandinavian origin, thus linking the deposit with the Corton Formation.

Corton till

The tills comprise diamicts of very silty sandy clay or clayey sand, commonly laminated, with a scatter of pebbles. The sand content varies, and fine sand laminae or lenses occur, commonly cross-bedded. The tills are consistently brownish grey to dusky yellowish brown in colour, but they are somewhat greyer in the north-west of the district ('Starston Till') than in the north-east ('Corton till' or 'Norwich Brickearth'). The tills are firm to stiff and stand in near-vertical faces when fresh, but in surface outcrops they are commonly decalcified, and they weather rapidly to a soft and friable condition as a result of their high sand content.

Mechanical analysis of the < 2 mm fraction of samples from the area of Beccles and Burgh St Peters (Hopson, 1991) showed the deposit to be extremely uniform and to comprise 22.0% clay, 23.6% silt, and 54.4% fine- and medium-grained sand. Pebbles and coarse sand (> 4 mm) account for only 4.7% by weight, and include a high percentage of coarse sand-grade chalk. Pebbles are mostly flints, both angular and rounded, with subordinate vein quartz, quartzite, chalk and shell fragments (derived from the Crag), and a sparse but diagnostic suite of rhomb porphyries, non-porphyritic lavas, mica schists, gneisses and granitic rocks believed to be of Scandinavian origin (Boswell, 1916). Analyses of Corton tills from boreholes in the north-east of the district (Hopson and Bridge, 1987) show the ratios of quartz, quartzite and flint to lie within those of the Bytham Sands and Gravels (Table 6; Figure 19c), with rather less flint or quartzite than the Corton gravels. The 4 to 8 mm size range comprises 54% flint, 24% vein quartz, 8% quartzite, 3% sandstone, and 2% each of limestone, ironstone and igneous/metamorphic rocks (Bridge and Hopson, 1985).

Chalk is commoner in the greyer tills ('Starston Till') of the north-west of the district than in those of the north-east; in the Diss district to the west, this till locally contains up to 50% of chalk pebbles (by weight), though on average chalk accounts for only 22.6% of the pebble content of the Corton till (Mathers et al., 1993). Pebble analysis was conducted on only one till sample in the north-west of the present district (borehole TM 27 NE/29 [2800 7962]), the 8 to 16 mm fraction comprising 51% angular flint, 27% rounded flint, 14% vein quartz, 7% quartzite, 1% other species, and only a trace of chalk. However, the bulk sample was described in the field as containing 'angular pebbles of flint and chalk in upper parts' so probably the chalk clasts were contained in the > 16 mm fraction.

Provenance

Boswell (1916) found evidence of derivation from the Crag, Eocene beds and Chalk as the ice-sheet crossed the North Sea basin, while Perrin et al. (1994) found the till to be low in opaque heavy minerals and high in sand content (the opposite of the till in the Lowestoft Till Formation), and with little or no material derived from Jurassic, Lower Cretaceous or Palaeogene strata. They concluded that whilst the ice had assimilated variable amounts of chalk, it had not crossed substantial outcrops of limonitic Jurassic or Lower Cretaceous formations, or Palaeogene or Mesozoic clays.

Bridge and Hopson (1985) found a heavy mineral suite similar to that of the upper part of the Crag, with a high incidence of unstable minerals (green metamorphic hornblende and epidote) compared with resistate species (zircon, rutile, tourmaline and kyanite) and concluded that large amounts of Crag were incorporated into the ice sheet as it crossed the North Sea basin. Hallsworth (1994) came to a different conclusion, however. She analysed four samples of sand from the present district, from Rigg [4082 8980], Barsham Hill [4002 8947], and Playters Old Farm (two samples) [4420 8770]. The sands are characterised by unstable minerals, amphibole, epidote and garnet being particularly common, but the amounts of these present, and the ratio of apatite to tour-

maline, are both higher than in the Crag, suggesting that the sands do not contain much Crag material reworked from the floor of the North Sea. On the contrary, the abundance of amphibole, epidote and garnet imply that the detritus is of first-cycle origin, and the garnet geochemistry suggests derivation from low- to medium-grade-metasedimentary rocks. This would accord with derivation from Scandinavia, although the Moine and Dalradian terrains of the Scottish landmass would yield similar results.

Depositional environments

Harmer (1902) considered that the Norwich Brickearth was formed in a glacial lake impounded by ice. Wilcox and Horton (1982) followed Harmer (1904) and Boswell (1916) in regarding the till as waterlaid, because it contains bedding traces. They regarded it as a flow till, and considered that both the tills and the sands of the Corton Formation were deposited in a lake that formed ahead of the advancing Scandinavian ice. Hopson and Bridge (1987) considered that the gravel and sand facies respectively represented proximal and distal outwash deposits of the Scandinavian ice, the sands being deposited in an extensive body of shallow water and not as a fluvioglacial sandar. The cyclicity within the sands they related to oscillations in the position of the ice margin. However, they regarded the tills as lodgement tills because of the lack of sorting which they would have expected in a water-laid deposit, even though they recorded laminations within the till at Corton. Arthurton et al. (1994) considered that the absence of subglacial piping and tunnel valley formation, which are associated with the Lowestoft Till Formation, argues against an origin as lodgement till.

Eyles et al. (1989) demonstrated that the close association of the tills and sands in the Corton Formation indicates closely allied depositional environments, with widespread subaqueous conditions. They considered that the tills were formed by the 'rain-out' of unsorted debris from floating ice, combined with downslope sediment gravity flow. The formation of the tills from floating ice accords well with the sedimentology and geometry of the tills known from the present district. The continuous sheet of till to the north in the Great Yarmouth district may have formed beneath a floating ice shelf, whereas the patchy till known in the present district may have formed by the melting of icebergs which had broken away from the ice shelf.

Eyles et al. (1989) considered the formation to be glaciomarine rather than glaciolacustrine in view of the abundant shells and shell fragments present, the great lateral extent of the tills, the coarse character and thickness of inferred shoreface and beach deposits and the geographical restriction of the sequence to the margin of the North Sea Basin. However, the shells may have been derived from earlier North Sea deposits over which the Scandinavian ice must have travelled, and the other features could also be found in a lake of large dimensions. It is now widely believed that the Dover Strait was cut by the overflow from a proglacial lake (Stamp, 1927; A J Smith, 1985, 1989), and this almost certainly happened during the Anglian period (Gibbard, 1988; Hamblin et al., 1992). It is suggested that a lake formed between the ice sheets advancing from the north and the Chalk escarpment in the south, fed by many large rivers including the Thames and Rhine; that this eventually overflowed into the English Channel, cutting a gorge through the Chalk and forming the Dover Strait; and that the subaqueous tills and sands of the Corton Formation were deposited within this lake.

Details

Hedenham–Bungay–Wortwell–Metfield–Chippenhall Green

Figure 21 shows the distribution of the Corton Formation and of till within it, as proved by boreholes for sand and gravel assessment in the western part of the district (Auton et al., 1985). Out of 126 boreholes sunk in the area shown, 50 proved the Corton Formation to be absent, 22 proved it to be present without till and 35 proved it to include till; the remaining 18 failed to reach the Corton Formation. The boreholes proved a complex sequence of sands, pebbly sands and till, although in all but three of the boreholes no more than one till was recorded. The thickest sequence is 16.2 m at borehole TM 28 SE/33 [2668 8432], which lies in the Diss district, where 13.9 m of sand and pebbly sand overlay 2.3 m of till. The thickest record of sand was 13.7 m in borehole TM 38 NW/43 [3351 8803], and of till 8.4 m in TM 28 SE/44 [2884 8017]. It may be assumed that the Corton Formation was originally deposited throughout the area shown in the figure and that where it is now absent it has been removed by later erosion, notably in the large north-east-trending area corresponding to the present-day Waveney valley. However, where the Corton Formation is shown to be present but not to include till, it is likely that till was never deposited.

Where the formation crops out in the valley sides of the Broome Beck around Woodton [29 94], it is impossible to distinguish it from the underlying Bytham Sands and Gravels in degraded pit sections. Chalky medium-grained sands of the formation were noted on the lower slopes of the north side of the Waveney valley at Holbrook Hill [283 863], but could not be separated from the Lowestoft Formation on the map; till was noted near Round House [2848 8652]. The Starston Till crops out on the interfluves of the Dentonwash tributary of the River Waveney near Denton [2888 8742, 2897 8750, 2942 8716], where it comprises a stiff brown chalk-free sandy clay. A borehole (TM28NE/3 [2787 8847]) at Chapel Farm, Denton, encountered 3 m of 'clay' at a depth of 23.2 m, and another borehole (TM 28 NE/11 [2865 8839]) at Hall Farm encountered 4.6 m of brown clay at a depth of 19.8 m, both probably Starston Till. Smooth grey clays were noted on the north side of the valley which runs towards Metfield [2799 8099, 2813 8091].

Horton (*in* Lawson, 1982) logged detailed sections and boreholes in the 'Beccles Beds' at Flixton Quarry [29 86], but unfortunately it is unclear which parts of the sections represent the Corton Formation; a typical example [2940 8611] in the Beccles Beds is as follows:

	Thickness m
Gravel, poorly sorted; coarse-grained sand matrix; sharp base	0.55
Clay, reddish brown to chocolate brown; silty, scattered pebbles; lensoid, planar bedding below, sharp base	0–0.75
Gravel, reddish; well-sorted, pebbles to 10 cm diameter, uneven erosive base	c.0.15
Gravel, brown; well-rounded and angular pebbles in sandy matrix	up to 0.30
Gravel, brown, poorly sorted; clayey sand matrix; uneven erosional base	0.35
Sand, pale brown, fine-grained, well-sorted	0.16
Clay, brown above, dark grey below	0.80
Sandy clay, reddish brown	0.05
Sand, pale brown, medium- to fine-grained, well-sorted	0.15+
Gravel, well-rounded clasts	—

The Corton Formation crops out extensively on the south side of the Waveney valley from Flixton to Bungay, although on the north side of the valley it is generally obscured by the Aldeby Sands and Gravels. A trial pit at Oaklands Farm, Flixton [3210 8755] recorded:

	Thickness m
LOWESTOFT TILL	0.45
CORTON FORMATION	
Sand, brown, slightly clayey	0.58
Clay, pale buff, soft, small chalk grains; diffuse passage base	0.02
Sand, pale yellow, chalk grains; passage base	0.80
Clay, pale buff, soft, chalky; diffuse passage base	0.05
Sand, pale yellow, well-sorted, rounded grains, chalk grains common	1.25

A section with more detrital chalk was measured in the disused pit south of Annis Hill [3509 8962]:

	Thickness m
LOWESTOFT TILL	1.56
CORTON FORMATION	
Silt, medium brown, slightly sandy to clayey, scattered small pebbles and chalk grains; passage base	0.28
Silt and laminated fine sand, medium brown	0.14
Clay, sandy to silty, abundant chalk pebbles 30–40 mm diameter, few quartzites; passage base	0.28
Sand, very silty, clayey, abundant chalk pebbles up to 50 mm diameter	0.30
Sand, yellow, fine- to medium-grained, slightly chalky	0.23+

The Starston Till was not found to crop out in the Bungay area but it was noted in boreholes, the thickest record being 12.5 m at All Hallows Home, Ditchingham (TM 39 SW/10 [3343 9230]) where it was overlain by 18.3 m of Corton sand. South of Bungay, sand and gravel assessment boreholes record sands and gravels but no till, though 0.9 m of 'brickearth' recorded in a borehole at Trinity Farm (TM 38 NE/14 [3548 8893]) suggests that it is present locally.

At Leet Hill on the northern flank of the Waveney valley, some 3 km downstream from the village of Broome, a distinctive sand and gravel unit ('Leet Hill Sands and Gravels', Hopson and Bridge, 1987), has been identified in quarry sections and boreholes. Although these could not be separately mapped, boreholes suggest they are quite widespread. At the time of survey, the pit at Leet Hill [3807 9294] only exposed 2.8 m of pale brown sand overlying 7.2 m of pale fawn to yellow sand, the latter being medium-grained, locally chalky, flat bedded with occasional rippled or cross-bedded units, and containing pebbly lenses with rare chalk and clayey silt pebbles. However, examination (Hopson, 1991) of new sections revealed

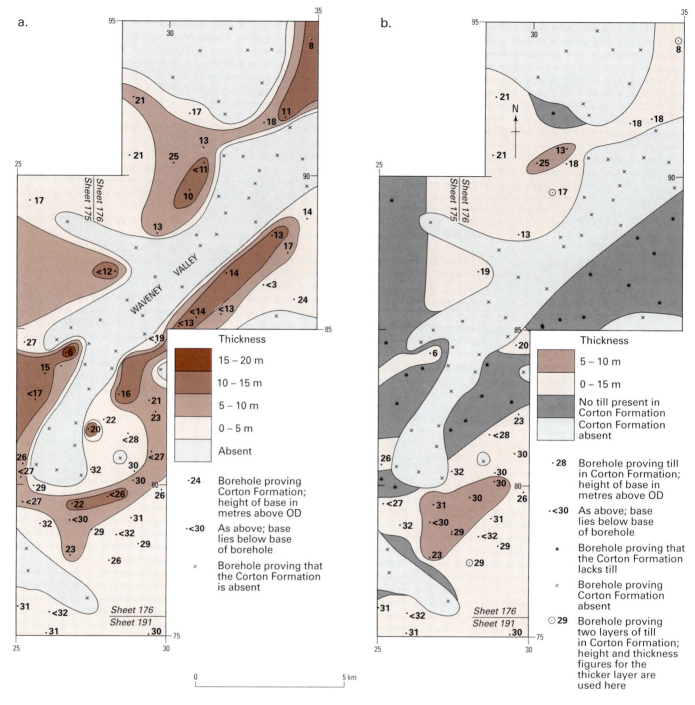

Figure 21 Variation of the Corton Formation proved in boreholes in the western part of the district.

a. Corton Formation (undivided): distribution, height of base and thickness.
b. Till in Corton Formation: distribution, height of base and thickness.

the Leet Hill Sands and Gravels at a deeper level, underlying the 'Starston Till' and overlying Bytham Sands and Gravels:

CORTON FORMATION

	Thickness m
Sand, pale yellow, fine- and medium-grained, chalky, rare pebbles; planar and ripple drift bedded	8.2
Sand, medium-grained, pebbly; purplish grey rhomb porphyries	0.6
Till; interlaminated brown silts, clays and thin sands with rare flint and quartz pebbles	0.7
Sandy gravel with medium- to coarse-grained sand lenses; cross-bedded; dominantly flint and quartz, cobble lag at base; Scandinavian erratics throughout in the fine gravel (4 to 8 mm) fraction	5.3

BYTHAM SANDS AND GRAVELS —

Since quarry working ceased, the lower unit has become obscured by talus. The Leet Hill Sands and Gravels here closely resemble the underlying Bytham Sands and Gravels, apart from the presence of the Scandinavian material.

Gillingham–Toft Monks–Burgh St Peter–Barnby

Sands of the Corton Formation crop out on the flanks of the Waveney valley, and gravels are known at outcrop within the sands at four localites [427 926, 468 949, 492 908, 486 931]; at the first two, fine- to medium-grained sands become pebbly downwards and rest upon pebbly sands and sandy gravels containing purple and grey rhomb porphyries. The only outcrop of till lies south-east of Beech Farm [486 931]. Seven of the thirty boreholes drilled in 1973 between Corton and Beccles (Hopson and Bridge, 1987) lie within the present district. These showed the base of the Corton Formation, resting directly on the Bytham Sands and Gravels, to rise generally westward from 0.1 m (TM 49 SE/8 [4644 9326]) to 6.7 m above OD (TM 49 SW/133 [4067 9489]). Four boreholes proved till, in all cases at the base of the formation, thickest (2.8 m) at Bull's Green (TM 49 SW/135 [4179 9423]). All the boreholes proved sands, in four cases passing down into pebbly sands and gravels. The thickness of sand and gravel ranged from 5.7 m (TM 49 SW/135) to 14.7 m (TM 49 SW/133). The eroded top of the sands, which in all the boreholes was overlain by Lowestoft Till Formation, rose westwards from 13.1 m (TM 49 SE/9 [4831 9303]) to 24.2 m above OD (TM 49 SW/134 [4009 9337]).

Beccles–Redisham–Kessingland–Lowestoft

Sands of the Corton Formation crop out on the southern flank of the Waveney valley, in the valley of the Hundred River and in the dry valley which runs from Mutford to Pakefield, with an extensive outcrop in Lowestoft itself. No till is known around Beccles or Redisham, but an outcrop has been mapped at Hulver Street [468 866 to 471 870]. Along the coast the sands are visible in the cliff from Kessingland [536 859] to Lowestoft. They are overlain by till of the Lowestoft Till Formation and underlain by Cromer Forest-bed Formation, but the base is generally obscured by landslip or coastal defences. The following section was measured at Crazy Mary's Hole [5366 8867]:

	Thickness m
LOWESTOFT TILL	2.0
CORTON FORMATION	
Sand, buff to yellow, soft, fine-grained, cross-bedded basal 2 m coarser, with diffuse layers of flint gravel: flints up to 7 cm long, angular and rounded, also quartz and quartzite pebbles; clear-cut base	7.0
CROMER FOREST-BED FORMATION	—

Blake (1884a) measured this section at a time when the cliffs were retreating and the base of the Corton Formation was exposed from Kessingland to a point about 360 m north of Crazy Mary's Hole. He showed the sands to be up to 8.8 m thick, with up to 3.4 m of pebbly sand beneath. The pebbly beds were present continuously north of Crazy Mary's Hole, but were absent for most of the distance south of there. Four small lenticles of till were noted: two to the south of Crazy Mary's Hole, up to 2.7 m thick and lying between Corton sand and Cromer Forest-bed Formation, and two smaller patches north of Crazy Mary's Hole both resting between sand and pebbly sand of the Corton Formation. Farther north no till is visible in

the cliff sections as far as Corton [5451 9722] in the Great Yarmouth district, where it forms a continuous sheet at least 3.3 m thick (Blake, 1884a), but it is known in boreholes in Lowestoft. These boreholes also indicate that the Corton sands are associated there with thick pebbly sands and gravels.

Sotterley–Benacre–Wrentham–Southwold

Sand of the Corton Formation, around 2 m thick, crops out widely beneath till of the Lowestoft Till Formation from Wrentham to Benacre and South Cove, and also as six outliers capping hills along the coast as far south as Easton Bavents [511 778]. The Covehithe cliffs expose up to 4 m of buff, largely homogenous, sporadically pebbly sand and very sandy till. Crude parallel bedding was noted, as were angular and rounded granules of chalk.

LOWESTOFT TILL FORMATION

Distribution and classification

The Lowestoft Till Formation is the most extensive surface deposit, forming a dissected, undulating sheet over much of the higher ground of the district. The most continuous outcrops are in the west, where only in the deeper valleys has erosion cut through to the base of the formation. In addition to capping the higher ground, the formation also occurs, in the north of the district, at lower levels within some of the major valleys. In the south-east the Lowestoft till infills elongate ice-marginal channels. Sands and gravels associated with the till have until recently been classified separately as Glacial Sand and Gravel or Glaciofluvial Deposits on BGS maps but on the Saxmundham and Lowestoft sheets these sands and gravels are included with the till as part of the Lowestoft Till Formation following the precedent set on the adjoining Great Yarmouth sheet (Arthurton et al., 1994). The term Lowestoft till is used to distinguish the till lithology within the formation. Local informal names have been given to some sands and gravels (Corton Woods Sands and Gravels, Haddiscoe Sands and Gravels, and Aldeby Sands and Gravels), and to some silts (Oulton Beds), where distinctive units can be recognised in the field (Figure 22).

The Lowestoft till has in previous accounts been referred to variously as the Lower Erratic Tertiaries — Chalky-Jurassic boulder clay of south Norfolk (Trimmer, 1851), the Upper Chalky or true Boulder Clay (Wood and Harmer, 1868), the Upper Glacial or Chalky Boulder Clay (Wood, 1880), Boulder Clay or Chalky Boulder Clay, (Woodward, 1881), the Chalky Boulder Clay (Blake, 1884; Harmer, 1902)), and more recently as the Lowestoft Boulder Clay (Baden-Powell, 1948) and as the Lowestoft Till (Banham, 1971, 1975, 1988; Bristow and Cox, 1973; Perrin et al., 1973; Hopson and Bridge, 1987; Rose, 1989) and Lowestoft Till Formation (Arthurton et al., 1994).

In the south of the district ground conductivity surveys have been used extensively to define the margins of the till sheet and of some adjacent melt-water channels (Figure 23; Mathers and Zalasiewicz, 1986). Elsewhere the limits of the till are generally reflected by changes in

Figure 22   Schematic cross-section of the Lowestoft Till and Corton formations in the north-east of the district, showing interrelationships and named and unnamed units described in the text.

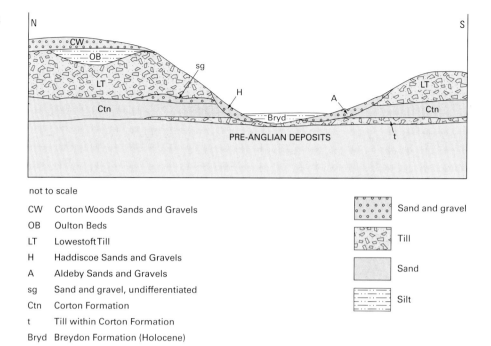

not to scale

CW   Corton Woods Sands and Gravels
OB   Oulton Beds
LT   Lowestoft Till
H   Haddiscoe Sands and Gravels
A   Aldeby Sands and Gravels
sg   Sand and gravel, undifferentiated
Ctn   Corton Formation
t   Till within Corton Formation
Bryd   Breydon Formation (Holocene)

Sand and gravel

Till

Sand

Silt

soil type and can be mapped without recourse to indirect methods.

Age and provenance

The provenance and relative ages of the chalky tills which cover large parts of eastern England Harmer, (1910, 1928) have been discussed by Hill (1902), Hollingworth and Taylor (1946), Rice (1968), Horton (1970), Perrin et al. (1973, 1979) and Hopson and Bridge (1987). Possible subdivisions of the Lowestoft till into a Lower Chalky Boulder Clay (Lowestoft till) and an Upper Chalky Boulder Clay (Gipping till) separated in time by the Hoxnian Interglacial has been discussed by Baden Powell (1948), West and Donner (1956), Straw (1960), West (1963) and Bristow and Cox (1973). The general uniformity of the chalky till over much of eastern England led Perrin et al. (1973) to conclude that it was the product a single glaciation, thus supporting the views of Bristow and Cox (1973), based largely on field evidence, that all the 'Chalky Boulder Clay' south of Norwich formed a single pre-Hoxnian sheet (the Lowestoft till), and that the Gipping till is a lithological variant of it. Sumbler (1983) has taken the argument further, arguing that the Wolstonian type sequence in Warwickshire and the Anglian type sequence in East Anglia were laid down during the same pre-Hoxnian stage.

In terms of the British Quaternary classification, the Lowestoft Till Formation belongs to the Anglian Stage (Mitchell et al., 1973). Within the present district the formation is overlain by Hoxnian interglacial deposits, and a similar relationship is found elsewhere, for example at Sicklesmere (West, 1981) and Icklingham (Kerney, 1976; Holyoak at al., 1983) to the west of the present district (Bristow, 1990). From oxygen isotope results on shells from ocean-floor sediments, the base of

the Hoxnian is placed at 250 000 BP (Shackleton and Opdyke, 1973). The age of the Lowestoft Till Formation in terms of glacial chronology recognised on the continent has been much discussed. Bristow and Cox (1973) concluded that it was dependant on the equivalence of the Hoxnian with the continental Holsteinian.

The provenance of the constituent material, coupled with directional measurements on the included pebble fabrics (West and Donner, 1956) and the regional trends observed within the composition of the matrix (Perrin et al., 1979), indicate that the Lowestoft till was the subglacial product of a mass of ice probably originating in the North Sea. One of the simpler explanations (Perrin et al., 1979) is that this ice stream moved southwards down the western edge of the North Sea Basin, accumulated chalk and oolite- and limonite-bearing Jurassic debris off the coast of Yorkshire, overran the Cretaceous escarpments in Lincolnshire and Norfolk, and excavated the Jurassic clays of the Wash and Fenland basin. Subsequently, the sheet fanned out radially with lobes spreading eastwards and south-eastwards, or according to Hart (1987) eastwards then northwards across the district.

Much of the Lowestoft till is probably a lodgement till (Woodward, 1881, p.115; Harmer 1910, Perrin et al., 1973 and Bridge and Hopson, 1985), although other processes of till formation, including deformation, flow or melt out are likely to have contributed.

Figure 23   Anglian meltwater channels (after Mathers and Zalaziewicz, 1985).

a. Conductivity contour map of the Aldeburgh–Snape area defining the location of glacial meltwater channels.
b. Interpretive cross-section and conductivity profile of a glacial meltwater channel at Snape Hall located on Figure 23a.

a.

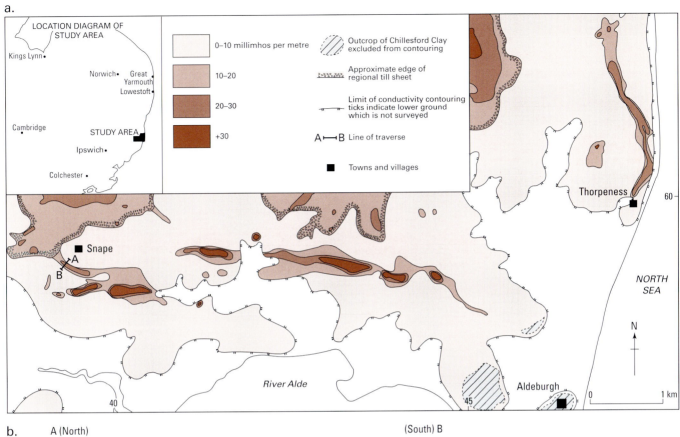

LOCATION DIAGRAM OF STUDY AREA

Kings Lynn •

Norwich •   Great Yarmouth •
            Lowestoft •

Cambridge •

STUDY AREA ■

Ipswich •

Colchester •

	0–10 millimhos per metre
	10–20
	20–30
	+30

Outcrop of Chillesford Clay excluded from contouring

Approximate edge of regional till sheet

Limit of conductivity contouring ticks indicate lower ground which is not surveyed

A ⊢━━┥ B   Line of traverse

■   Towns and villages

■ Snape

A

B

Thorpeness ■

60

NORTH SEA

N

River Alde

Aldeburgh ■

45

40

0   1 km

b.

A (North)   (South) B

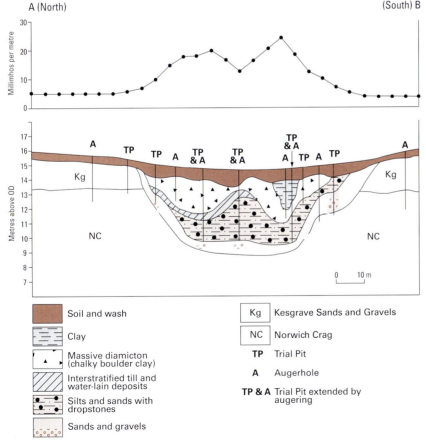

30

20

10

0

Millimhos per metre

17
16
15
14
13
12
11
10
9
8
7

Metres above OD

A   TP   TP   A   TP   TP   TP   A TP A TP   A
                            &A   &A   &A

Kg   Kg

NC   NC

0   10 m

	Soil and wash
	Clay
	Massive diamicton (chalky boulder clay)
	Interstratified till and water-lain deposits
	Silts and sands with dropstones
	Sands and gravels

Kg   Kesgrave Sands and Gravels

NC   Norwich Crag

TP   Trial Pit

A   Augerhole

TP & A   Trial Pit extended by augering

Lithostratigraphy

The till is a stiff, bluish grey, chalky, flinty, variably silty and sandy clay that weathers yellowish brown. Other pebbles within the clay include quartzite, vein quartz, iron-pan, and Jurassic material such as limestone, mudstone, cementstone and derived fossils. In addition to the chalk clasts much fine-grained chalk is present within the clay matrix. Patches of silt, generally with only minor quantities of pebbles, occur within or (more commonly) at the base of the till and are locally mappable. They are considered to be glaciolacustrine in origin. Lenses of sand or gravelly sand up to several tens of metres across (known locally to farmers as 'gaults', 'galls' or 'golts') occur within or above the till. These sandy patches commonly become waterlogged following prolonged rain and can pose problems to heavy farm machinery. Removal of these patches by farmers accounts for some of the shallow depressions present in the fields. Such deposits are too small to show on the map.

Within the basal metre or so, the till is commonly brown and is either chalk-free or contains only scattered pellets of chalk. The proportion of clay in the matrix is also lower than in the succeeding beds, whilst the amount of sand and gravel is proportionally increased, probably reflecting incorporation of the underlying sands and gravels. Crude lamination is locally present, which suggests deposition in a viscous or fluid state, possibly as a flow till. It is probable that some of the sands and gravels mapped locally within the formation may be this basal part of the till sheet. It has been suggested by Ellison and Lake (1986) that the lack of chalk in the lower part of the till may be a primary depositional feature, related to the increased solubility of calcium carbonate in cold water; chalk debris in the base of the icesheet may have been removed prior to deposition of the basal till.

The mineralogy and lithology of the Lowestoft till has been examined by Perrin et al. (1973). From their work they concluded that the till is very uniform from Lincolnshire to Hertfordshire, although there are minor differences in the fine-grained sand and silt fractions between the extreme north and south of this region. The average calcium carbonate content increases southwards from 23 to 43%, and this is believed to be due to the incorporation of increasing quantities of chalk by the ice as it moved southwards. The clay mineralogy shows no significant differences with geographical location; the assemblages being of mica and kaolinite, with variable smectite and locally a little chlorite. This uniformity makes it impossible to determine which of the local clays in the Jurassic or Cretaceous succession has provided the dominant source of the clay minerals. Stone counts show chalk (55 to 84%) to be the dominant clast type, with flints and 'others' (mostly Jurassic) in roughly equal proportions. Far travelled 'exotics' account for only 0.1% of the total.

Analyses of the derived fossil content (Woodward, 1881, p.115; Blake, 1890, p.53) indicate that amongst the more common species observed are *Gryphea incurva* (Lower Lias), *Gryphea dilitata* (Oxford Clay); ammonites from the Lias, Kimmeridge Clay and Oxford Clay; and *Bellemnites abbreviatus* from the Corallian and Oxford Clay.

Marginal meltwater channels

Several infilled glacial meltwater channels are present near Snape and Thorpeness in the southern part of the district. The extent and geometry of the channels (Figure 23) have been investigated by a combination of field survey techniques including augering and conductivity surveys using a Geonics EM31 conductivity meter (Mathers and Zalasiewicz, 1986; Cornwell and McCann, 1991).

The channels, up to 4 km long by 150 m wide and 8 m deep, show a consistent pattern of infill, comprising glaciofluvial sands and gravels overlain by subglacial waterlaid fine-grained sands and silts; these pass conformably up through thinly interstratified tills and sands into the basal part of the till sheet.

The absence of pre-Anglian deposits within the channels and the presence at the base of the channel sequences of glaciofluvial sediments indicates that they were cut by meltwater issuing from the Anglian icesheet. The relatively straight alignment of the channel systems suggests controlled rather than free drainage. The regional distribution of Anglian deposits shows that the channels were initiated close to the maximum extent of the ice sheet. Rare occurrences of till in the basal glaciofluvial sediments indicate the proximity of ice during the early stages of infill. The channels are interpreted by Mathers and Zalasiewicz (1986) as erosional furrows cut parallel to the ice margin and then overridden by ice and infilled subglacially.

It is suggested below that the Oulton Beds, the Corton Woods Sands and Gravels, the Haddiscoe Sands and Gravels and the Aldeby Sands and Gravels, in the north of the district, also formed within ice-marginal channels. However, these are considered to have formed at stillstands during the wastage of the 'Lowestoft' ice sheet rather than during a period of ice advance, because they are underlain, not overlain, by the main Lowestoft till sheet (Figure 22).

Oulton Beds

Nomenclature and distribution

Oulton Beds are proved only in boreholes in the north and north-eastern parts of the district (see Figure 20) where they occur between the Lowestoft till and the Corton Woods Sands and Gravels (Figure 22).

The Oulton Beds were defined by Banham (1971) in cliff exposures at the Anglian type section [546 970 to 548 965], north of the present district, where they occur between the Pleasure Gardens Till and Lowestoft till. The term Oulton Beds is derived from the village of Oulton where possible lateral equivalents of these sediments were noted by Blake (1890, pp.56–57 and pp.60–61); these occurrences have been not confirmed by BGS surveys.

Within the district, the Oulton Beds consist of laminated grey clay up to 3 m thick with intercalated silts and sands. Banham (1971) recorded a maximum thickness of 4 m. The lower part of the sequence, up to 2.5 m thick, is a stiff grey clay and this passes up into interlaminated silt and sand which is characteristically grey to buff in colour). The deposit as a whole has a variable content of chalk, small flint pebbles and carbonaceous material. There is general agreement that the Oulton Beds are glaciolacustrine in origin (Banham, 1971; Pointon, 1978; Bridge and Hopson, 1985).

Corton Woods Sands and Gravels

Nomenclature and distribution

Corton Woods Sands and Gravels were formerly mapped as 'Gravel, Sand and Loam' on the Old Series one-inch map (Sheet 67 SW, 1882) and referred to in the Yarmouth and Lowestoft memoir (Blake, 1890) as 'Plateau Gravels'. The deposits have a restricted occurrence in the district; they crop out north of Oulton Broad and Lake Lothing, generally above the 15 m OD contour although locally they occur as low as 10 m OD [as at 527 933]. An isolated low-level outcrop [centred on 5160 9350] near Lowestoft may alternatively be the lateral equivalent of 'Leet Hill Sands and Gravels' (see Corton Formation).

Stratigraphy and sedimentation

Within the district the deposit generally overlies Lowestoft till, although locally it rests on the Oulton Beds or Corton sands (Figure 22). Further north at the Corton Woods type locality [5467 9653] they overlie the Pleasure Gardens Till and cap the Anglian succession. Mapping suggests that the deposit is a layer typically 2 to 3 m thick which drapes downslope at the outcrop margins; thicknesses in excess of 5m are recorded in a few boreholes and at one poorly exposed section [5455 9415].

The Corton Woods Sands and Gravels are distinguished by strong reddish brown to orange-brown colours. The gravel component, mostly of medium gravel (8 to 16 mm), ranges from angular to subangular to pebbles, and comprises brown 'weathered' flints (about 66%), with subordinate quartz (about 18%) and quartzite (about 7%) (Bridge, 1993). Arthurton et al. (1994) recorded a strong imbrication in the gravels, but this is not evident in the few exposures in the district. The sand fraction ranges from medium to coarse (0.25 to 1.00 mm).

Pointon (1978) interpreted this suite of sands and gravels in the upper part of the Corton section as outwash from the 'Lowestoft' icesheet, and stratigraphically conformable with the underlying deposits. However, work by Bridge and Hopson (1985) suggests that the Corton Woods Sands and Gravels have no mineralogical affinity with the Lowestoft till, and that they rest on an erosion surface cut across the underlying Anglian deposits. These authors concluded that the Corton Woods Sands and Gravels were deposited under periglacial conditions imposed by a glaciation later than the Anglian. However, if the gravels postdate the Anglian

glaciation it is difficult to envisage a mechanism by which they could have been deposited on the plateau tops, as such areas would have been surrounded by valley systems.

Details

Lowestoft–Normanston–Oulton

Although the Corton Woods Sands and Gravels outcrop over approximately 3 km² within this area, there are few exposures and only one of these presents a clear section. Detail on thickness and distribution is from borehole information and hand augering.

In a disused railway cutting [5455 9415] up to 8 m of dark orange sand and angular to subangular gravel overlie Lowestoft till. About 2 m of Corton Woods Sands and Gravels overlying Lowestoft Till are exposed in a disused pit [528 934]. Corton Formation is exposed below. In another pit [5112 9430] the following section was recorded:-

	Thickness m
CORTON WOODS SANDS AND GRAVELS	
Pebbly topsoil	0–0.4
Sand and angular to subangular gravel	0.4–1.5
CORTON FORMATION	
Sand, fine, silty	1.5–8.0

At Camps Heath disused gravel pit [518 946] a section at the eastern end shows up to 4 m of Corton Woods Sands and Gravels comprising dark orange sand with angular to subangular gravel comprising mainly flint with occasional subrounded quartzite.

Haddiscoe Sands and Gravels

The Haddiscoe Sands and Gravels (Arthurton et al., 1994) are restricted in the present district to a small area along the south side of the Waveney valley [between 450 950 and 490 930] to the west of Lowestoft. They were referred to as the Valley Flank Sands and Gravels by Bridge and Hopson (1985), cropping out along the valley sides up to 20 m above OD but also extending locally beneath the Holocene valley floor deposits.

The type section is at Haddiscoe Pit [445 963], in the adjacent Great Yarmouth district, where a black, well-rounded fine to coarse flint gravel overlies sands of the Corton Formation. Within the gravels large rotated blocks of sand are preserved, which Bridge and Hopson (1985) believe could only have been incorporated in a frozen condition. Mapping indicates that the gravels overlie and cut into the Lowestoft till, implying that their age is late Anglian or younger (Arthurton et al., 1994). However, the sequence in the Haddiscoe area is not simple, as in the Haddiscoe Old Pit [444 966] Hey (1967, pp.437–438) recorded apparently similar rounded gravels beneath Lowestoft till and Corton Formation. He considered these to be of post-Baventian age, an outlier of the 'Westleton Beds' (now regarded as part of the Crag Group). Thus at Haddiscoe two gravels of similar lithology, but of very different age, are present.

Heavy mineral analysis (Hopson, 1991) reveals that the Haddiscoe Sands and Gravels are low in opaque iron minerals and high in resistate species such as zircon and rutile. They appear to be unrelated to the Lowestoft Till Formation, therefore.

Details

At Tom's Carr Pit [4720 9468] the following section was recorded by Hopson (1991):

Surface level c. 24 m above OD

	Thickness m
HADDISCOE SANDS AND GRAVELS	
Yellow-brown sand, medium to coarse with some fine; planar-bedded fine flint gravel stringers	1.5
Flint gravel up to cobble grade, with very well-rounded interstitial medium and coarse sand. Cross-bedding indicates current direction approximately towards SE	4.0

At Marsh Farm Pit [4765 9440] about 5 m of medium- to coarse-grained and well-rounded flint gravel were seen in intermittent exposures in overgrown and degraded workings.

Aldeby Sands and Gravels

These deposits, which postdate the main till sheet within the Lowestoft Till Formation, are restricted to the north-east of the district, where they extend in an east–west train along the valley of the Broome Beck, a tributary of the Waveney, then along the Waveney valley almost to the coast. The type section is the Oaklands Pit at Aldeby [4645 9290] where up to 18 m have been recorded (Hopson, 1991) (see details below). Similar sands and gravels are present along the valley of the Hundred River to the south of the Waveney valley.

In the valley of the Broome Beck the sands and gravels were originally considered to underlie the Lowestoft till of the plateau, but boreholes drilled during a sand and gravel assessment study (Auton et al., 1985) have proved that the sands and gravels are confined to the valley and are not present beneath the till along the valley flanks (Figure 22). For this reason the deposits that were classified on the 1:10 000 scale maps as 'glacial sands and gravels' and 'glacial deposits, undifferentiated' have been reclassified here, and on the published 1:50 000 map (Lowestoft Sheet), as Aldeby Sands and Gravels.

The Aldeby Sands and Gravels overlie Lowestoft till in places and the sands of the underlying Corton Formation elsewhere. Locally, for example at Stanley Hills [434 932], at Aldeby Hall [445 928], at Eastend Farm [496 921] and north of Worlingham Hall [443 902] the sands and gravels can be traced down to, and beneath, the peats and silts of the Breydon Formation.

The Aldeby Sands and Gravels are lithologically very variable; though they are predominantly sands and gravels they also contain lenses of reworked chalky till, discrete beds of clay and silt, reconstituted chalk, and masses of what appears to be transported sand from the Corton Formation. The gravel fraction is composed predominantly of clasts of angular flint with some quartz and quartzite. In places beds of well-rounded flint and quartz

gravel are present. Preliminary results of heavy mineral analysis indicate a strong affinity with the Lowestoft till (Hopson, 1991). According to Hopson and Bridge (1987) they are lithologically similar to the High Level Gravels of the middle Waveney (Coxon, 1979), which were deposited during the wastage of Lowestoft ice from the higher ground.

Field relationships indicate that the Aldeby Sands and Gravels postdate the deposition of the main chalky Lowestoft till of the plateau, and are themselves overlain by terrace sands and gravels, some of which are associated with Hoxnian and Ipswichian deposits.

Details

Broome Beck area

Sands and gravels have been mapped extensively along the valley of the Broome Beck. To the east of Spinks Hill the deposits were mapped at the 1:10 000 scale as 'Glacial Drift, Undifferentiated' but have been reclassified for the 1:50 000 map as Aldeby Sands and Gravels. In this area the sands and gravels are intermixed with lenses and layers of chalky till.

Aldeby area

The type section for the Aldeby Sands and Gravels (Hopson, 1991) is at the Oaklands Pit [464 926]. A generalised section is given below:

Surface level c.15 m above OD

	Thickness m
ALDEBY SANDS AND GRAVELS	
Gravel, sandy, clayey, cryoturbated, brown; gravel up to cobble grade with patches of sand and clean sandy gravel. In places the deposit has a distinct red and grey mottling suggesting development of a palaeosol	0–2.0
Gravels, yellow-brown, up to cobble grade, interbedded with medium to coarse sands with seams and lenses of chalky gravel and silty clays. Generally planar base	up to 10.0
Gravel, fine to medium, and coarse yellow sand with seams of fine and coarse gravel	up to 8.0

Boreholes in the base of the pit show that the Aldeby Sands and Gravels are underlain by Lowestoft till, and at greater depth by shelly Crag.

To the east of Oaklands Pit a number of disused, heavily overgrown and degraded pits can be seen at Boon's Heath [470 926]. These pits vary in depth from 4.0 to 8.0 m and reveal interbedded yellow-brown sands and gravels similar to those in the Oaklands Pit. The upper part of the deposit is heavily iron-stained and in parts indurated to form a hard caprock, debris from which abounds in the vicinity.

At Waterloo Pit [4261 9320] the following section was measured:

Surface level c.17 m above OD

	Thickness m
Topsoil	0.2
ALDEBY SANDS AND GRAVELS	
Flint, gravels, fine and coarse, angular, with thin sand stringers, cross-bedded	2.7

Till, reworked, pale bluish grey with chalk and flint pebbles, fragments of calcrete derived from the Corton Formation at base. Chalk pebbles streaked out in places; some faint laminations picked out by pale greyish brown colour	1.7
Flint and quartz gravel, coarse, angular, with a matrix of coarse angular flint sand	0.3
Interbedded, sands and gravels, cross-bedded and chanelled, some planar-bedded sand seams	4.9
Flint gravel, fine with some coarse, very heavily iron-stained, cross-bedded throughout	1.4

The deposits seen at Waterloo Pit exhibit complex inter-relationships. A few metres to the south-west of the above section the 2.7 m till becomes much attenuated and involved with low angle thrusting in the underlying gravels. The basal 1.4 m iron-stained gravel occurs in pockets cut into a blue-grey stiff chalky till apparently of lodgement type. Elsewhere in the pit seams of chalky gravel, finely laminated silt, reconstituted chalk and transported blocks of sands from the Corton Formation, with bedding preserved, have all been recorded (Hopson, 1991).

Lowestoft area

In the Lowestoft area the Aldeby Sands and Gravels are represented by thin spreads (generally up to 1.5 m thick) of coarse-grained sands and angular to subangular gravels overlying sands of the Corton Formation. Borehole TM59SW/251 [527 908] proved a thickness of 3 m.

Origin and relationship of the Oulton Beds, Corton Woods Sands and Gravels, Haddiscoe Sands and Gravels and the Aldeby Sands and Gravels

The distribution of the Aldeby Sands and Gravels along the Broome Beck–Waveney valley suggests that they were deposited within a channel system. The channel is interpreted here as an ice-marginal feature cut in late Anglian times during a still-stand in the local southwards decay of the ice front. The lenses and irregular bodies of chalky till within the sands and gravels almost certainly formed from pre-existing till which slumped, flowed or slipped into the valley from the adjacent plateau. The small areas of mapped silt associated with the sands and gravels probably result from the local ponding caused by the presence of the slumped till on the valley floor.

It seems likely that the Corton Woods Sands and Gravels, and the associated Oulton Beds silts, formed in a similar fashion during an earlier still-stand when the ice margin was farther to the north, and the degree of erosion less. The Haddiscoe Sands and Gravels may have formed in a similar environment, at a time intermediate between the deposition of the Corton Woods Sands and Gravels and the Aldeby Sands and Gravels.

The absence of a heavy mineral assemblage typical of the Lowestoft Till Formation in the Corton Woods Sands and Gravels and the Haddiscoe Sands and Gravels (Bridge and Hopson, 1985) can be explained if the material being excavated by the meltwater was composed predominantly of Crag material or Bytham Sands and Gravels; both of these could have been available upstream, either in situ or reworked from within the Lowestoft Till Formation sands and gravels.

Unnamed sands and gravels

Apart from the informally named sand and gravels mentioned above, there are widespread outcrops of unnamed sand and gravel, below, within and above the Lowestoft till. These comprise predominantly angular to rounded patinated flints in a sandy matrix. Chalk clasts are present where the gravels have not been decalcified. Locally the gravels have incorporated material from the underlying Crag, the Kesgrave Sands and Gravels or the Bytham Sands and Gravels, producing deposits of hybrid appearance.

INTERGLACIAL DEPOSITS

Deposits attributed to the Hoxnian and Ipswichian interglacial periods have been proved in small temporary excavations only, and therefore do not appear on the map, but for completeness they have been included in the side-margin of the published 1:50 000 map (Lowestoft sheet).

Hoxnian interglacial deposits

Just west of the present district, dark grey humic silty clays of lacustrine origin were formerly dug from the brick-pit [222 710] at Athelington. The pit was disused at the time of survey and there were no sections, but hand augering around the worked-out area proved that the interglacial sediments do not extend far beyond the margins of the pit. Coxon (1979) drilled a series of boreholes which showed that the clays were deposited in a closed depression upon the surface of the Lowestoft till. They rest directly upon the till, suggesting that the depression formed immediately after the retreat of the ice sheet and probably as a result of the melting of a mass of clean dead ice (Horton, 1982).

Coxon (1979) demonstrated that the sequence at Athelington can be correlated with the type site of the Hoxnian interglacial at Hoxne, some 9 km to the north-west, and with similar deposits at Saint Cross South Elmham, within the present district.

The interglacial deposits at Saint Cross South Elmham are situated on the south side of The Beck, a small tributary of the River Waveney, some 2.5 km from the river itself and 3 km east of Wortwell. The site [303 840] was originally described by Candler (1889) and was further investigated by West (1961b). Candler noted that the interglacial deposits occupy a ridge or tongue of land between depressions.

At Saint Cross South Elmham a variety of lithologies were recorded by Coxon (*in* Allen, 1984, pp.95–100). He described the lower sediments as dark grey to olive-green clay-muds deposited in a quiet low energy lacustrine environment, with sandy horizons indicating higher energy deposition. At higher levels the clay is mottled and contains greater amounts of sand. The top of the deposits is marked by an erosion surface, above which solifluced chalky till from the adjacent slopes has cut into the lake deposits. Boreholes show that the interglacial deposits have a maximum thickness of 11 m and lie in a basin some 300 m in diameter. Coxon suggested that the basin probably formed as a kettle hole in the Lowestoft till.

Pollen from the borehole samples has yielded the following results:

Local pollen assemblage biozones (p.a.b.)	Regional correlation with Marks Tey Turner, 1970) and Hoxne (West, 1956)
6. *Alnus-Corylus* p.a.b.	Ho IIc3
5. *Alnus*-Gramineae-*Corylus* p.a.b.	Ho IIc2
4. *Alnus-Corylus-Quercus* p.a.b.	Ho IIc1
3. *Alnus-Quercus* p.a.b.	Ho IIb
2. *Betula-Pinus-Quercus* p.a.b.	Ho IIa
1. *Betula*-Gramineae p.a.b.	Ho I

Type X pollen is present throughout the pollen record at Saint Cross South Elmham. The high non-aboreal pollen value in IIc is a widespread occurrence in Hoxnian deposits caused, according to Turner (1970), by a regional forest fire or heavy grazing by animals.

The Hoxnian sites along the Waveney valley lie on the margins of small tributaries that have cut down beside them. The proximity of the sites to the streams led Coxon (*in* Allen, 1984, p.100) to suggest that the basins might have been linked to the drainage system during their infilling. He noted that the degree of post-Hoxnian erosion marked by these tributaries is large for the area, around 10 m at South Elmham and Hoxne, and 5 m at Athelington, and that if the Hoxnian sites were linked to the drainage in the Waveney then presumably the streams were graded to a higher level in the main valley than at present.

Ipswichian interglacial deposits

Bones of an elephant (*Elephas antiquus*, Norwich Castle Museum No. 877.967) were found within sandy silty clay and silty mud in an excavation at Low Street, Wortwell [2752 8437]. Sparks and West (1968) obtained pollen and samples of plant macrofossils and freshwater molluscs from this deposit and assigned an Ipswichian interglacial age. Coxon (1979, pp.94–109) carried out a more detailed investigation, using augerholes, and showed that the organic sediments occur between 6.25 m and 14.5 m above OD, form a drape on a residual 'Homersfield Terrace' deposit, and are overlain non-sequentially by Head and sands and gravels of the first river terrace.

The eroded remnant at Wortwell is attributed to pollen zones lWo (Late Wolstonian) to Ipla-Ipllb (Ipswichian) and Ipllb (temperate Ipswichian). Similarly, the freshwater molluscs from one of Coxon's boreholes (Borehole No 1), at depths between 2.6 and 3.2 m and between 3.6 and 4.2 m, can be correlated with the Ipswichian type sequence at Bobbitshole, Suffolk.

The sediments at Wortwell are interpreted by Coxon (1979) as having been deposited in a low energy fluvial backwater. They are the only known Ipswichian interglacial sediments within the Waveney valley and demonstrate that the major downcutting of the valley had already taken place before the start of the Ipswichian Stage.

RIVER TERRACE DEPOSITS

River terrace deposits have been mapped along the valleys of the two major easterly flowing rivers in the district, the River Waveney and the River Blyth. In both valley systems the terrace deposits are present only in the west of the district; farther downstream they are absent. The eastern limit of the two higher terraces may reflect the tidal limit of the rivers during the periods of terrace aggradation. Within the Waveney valley the deposits of the lowest terrace can be demonstrated to extend beneath the level of the Holocene clays, silts and peats that occupy the valley floor. A similar relationship probably occurs in the Blyth valley but cannot be proven. Limited terrace deposits are also present along the valley of the River Deben in the extreme south-west of the district. No terrace deposits have been mapped along the valley of the Alde, although gravels below the Holocene alluvial deposits may represent buried deposits of the first terrace.

Waveney valley

Three river terraces have been recognised within this part of the Waveney valley (Figure 24). The third (Homersfield) terrace is the oldest and highest, forming a distinct but somewhat irregular topographical bench rising typically to about 6 m above the present floodplain. Its maximum height above the floodplain locally reaches several metres higher.

The second (Broome) terrace forms a bench at 2 to 4 m above the floodplain. Correlation of the terraces within the Waveney/Little Ouse catchment is complicated by the fact that in the adjacent Diss district to the west, two leaves of the Broome Terrace have been numbered separately, as second and third terraces, the Holmersfield Terrace being regarded there as the fourth terrace (Mathers et al., 1993).

The first (Floodplain) terrace, the most extensive of the terraces, forms intermittent low benches and mounds within the floodplain and rises to a height of about 1 m above the modern alluvium.

Figure 24a shows a long profile for the River Waveney between Wortwell and Geldeston. All the terrace deposits consist of sand and gravel: the gravel fraction is composed predominantly of subangular to subrounded flint with subordinate amounts of quartz and quartzite.

THIRD (HOMERSFIELD) TERRACE DEPOSITS

The Homersfield Terrace was named by Sparks and West (1968) after the village at the south-western end of a bench feature which extends between Homersfield [285 855] and Flixton [310 870]. Other significant outcrops within the district occur near Wortwell, to the west of Earsham, and at Bungay. Deposits of the Homersfield Terrace have not been recognised downstream of Bungay.

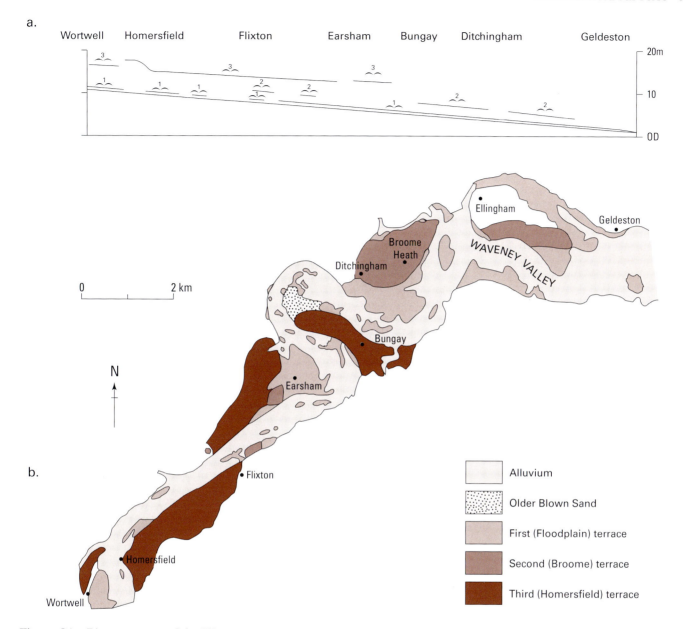

Figure 24   River terraces of the Waveney.

a. Long profile of River Waveney showing distribution of river terrace deposits between Wortwell and Geldeston.

b. Map of terraces in the same area.

Sections in the large quarry [295 862] at Flixton have been described by Horton (*in* Lawson, 1982), and by Coxon (*in* Allen, 1994, pp.80–86). Funnell (1955) recorded sections in a large now disused pit [287 857] at Homersfield and in another pit [2903 8568] at the south-west end of Flixton Park.

Coxon (1979, and *in* Allen,1984) regarded the deposits in the Flixton quarry as a thick complex sequence of fluvially deposited sands and gravels, with thin chalky clays which he interpreted as flow tills. He cited the existence of ice-collapse structures and a variable downstream terrace gradient as further evidence for the proximity of ice during aggradation and

suggested a late Anglian fluvioglacial origin for the deposits.

Subsequent investigations by Horton (*in* Lawson, 1982) and Clarke et al. (*in* Allen, 1984, pp.86–88) revealed an extensive silty clay within the gravel sequence. The clay is up to 2.5 m thick, dark grey to dark brown, and becomes orange-brown at its contact with the adjacent beds. Palynological examination of the clay has yielded disappointing results (Coxon, *in* Allen, 1984), as pollen is almost absent. Clarke et al. (*in* Allen, 1984) considered the clay to be stratigraphically very important as it separates gravels of very different facies. They believed that the terrace deposits are restricted to the gravel, up to

Table 8 Composition of the two gravels at Flixton Quarry (Clarke et al., *in* Allen, 1984).

Sample number	flint	quartzite	quartz	chalk	others
Upper gravels					
Sample Ton7	81*	14	5	–	trace
Sample Ton8	75	20	5	–	trace
Sample Ton9	74	15	8	1	2
Lower gravels					
Sample Ton1	59	31	9	–	1
Sample Ton2	60	35	5	–	trace
Sample Ton3	47	45	8	–	trace
Sample Ton6	53	33	11	–	3

*percentage by *weight* of the 8 to 16 mm fraction

4.5 m thick, above the clay. The lower gravel, beneath the clay, is up to 3 m thick and is characterised by a high proportion of well-rounded quartzite pebbles. This gravel overlies green shelly Crag. Analysis of the pebbles shows (Table 8) that the upper gravel contains a higher proportion of angular and subangular flints, and less quartz and quartzite, than the gravel below the bed of clay.

Clarke et al. (*in* Allen, 1984) regarded the lower gravel as fluvial in origin, but part of the pre-Anglian terraces of the River Thames (Kesgrave Sands and Gravels). Subsequent work has shown that these gravels are probably related to a more northerly river than the Thames, so would be more appropriately classified as Bytham Sands and Gravels.

Although many samples from the Flixton pit have proved to be barren of plant fossils (Coxon, *in* Allen, 1984), a sample collected by Funnell in 1954 was found to contain both micro- and macro-fossils. The major elements of the pollen assemblage from this sample (Coxon, *in* Allen,1984) are as follows:

Picea	3%
Pinus	10%
Betula	11%
Salix	3%
Gramineae	22%
Cyperaceae	24%
Caryophyllaceae	17%
Others	10%

Mammalian remains from the upper gravels at Flixton have been reported by Funnell (1955). They include *Mammuthus (Elephas) primigenius*, *Coelodonta (Diceros) antiquitatus*, *Megaloceros (Megaceros) giganteus*, and *Bos primigenius*. Stuart (*in* Allen, 1984, pp.84–86) identified further bones held in the Norwich Castle Museum as *Mammuthus primigenius*, *Coelodonta antiquitatis*, *Equus caballus*, *Rangifer tarandus*, *Bison* sp. or *Bos* sp.

Both the flora and fauna indicate a cold period of deposition, and imply a post-Anglian age.

SECOND (BROOME) TERRACE DEPOSITS

Within the present district deposits of the second terrace are restricted to two small outcrops and two larger spreads between Flixton and Geldeston (Figure 24b). It is probable that sands and gravels of the Broome Terrace also underlie the area of blown sand within the large meander at Outney Common, to the north of Bungay.

Upstream from Flixton the river valley is relatively narrow, and it is probable that any second terrace deposits have been destroyed by later downcutting of the river.

The second terrace was named the Broome Terrace by Sparks and West (1968) after the largest spread, and the only major exposures, at Broome Heath Pit [348 915] some 2 km north-east of Bungay. Sand and silt lenses within the sands and gravels contain pollen (Coxon, *in* Allen, 1984). This includes the Type X palynomorph, characteristic of the Hoxnian interglacial, which is present in small quantities as reworked material. As the terrace lies at a slightly higher topographical level than the nearby Ipswichian interglacial deposits at Wortwell, it was suggested by Coxon (*in* Allen, 1984) that the Broome Terrace deposits are Wolstonian in age.

At Broome Heath Pit, Coxon (*in* Allen, 1984) recognised three main units within the terrace deposits, in descending order:

a. cross-bedded sand with gravel and clay lenses

b. coarse, poorly sorted, bedded gravel and sand

c. sorted, cross-bedded gravel with sand

The upper and middle units contain both penecontemporaneous and post-depositional ice-wedge casts, and are also disturbed by cryoturbation in places (Coxon *in* Allen, 1984). The same author reported the average composition or the Broome Terrace deposits as flint 71%, quartz 14%, quartzite 10%, others 5%. Stuart (*in* Allen, 1984, p.91) has identified 13 molars and fragments of pelvis of *Mammuthus primigenius* from this pit, curated at the Norwich Castle Museum. The museum catalogue also records *Bison*.

The Broome Heath Pit has yielded an arctic flora (compilation by Coxon, *in* Allen, 1984). Herb pollen predominates, with Gramineae and Cyperaceae reaching 23% and 45% respectively. The tree and shrub taxa are diverse. The palynomorph Type X, restricted to the Hoxnian Interglacial (Turner, 1970; Phillips, 1976) is present at levels up to 2%, but was considered by Coxon to have been reworked. No macrofloral remains were present. A flint flake was found by Coxon 3 m below the surface in unit 'b'. It has been identified by Wymer (*in* Coxon, 1984) as a primary flake of black flint in a rolled condition.

The presence of ice-wedge casts, herbs, and *Betula* and *Pinus* in moderate frequencies suggests a cold depositional climate (Coxon, *in* Allen, 1984).

FIRST FLOODPLAIN TERRACE DEPOSITS

Deposits of the first terrace occur both as flanking benches, and as low 'islands' within the modern floodplain alluvium of the River Waveney (Figure 24b) from the extreme west of the district downstream as far as Geldeston. The 'islands' generally consist of fine- to medium-grained sands which become appreciably gravelly below 1 m depth. Along the valley sides the deposits are also sandy, but gravel is locally present at the surface, as for example in the area around the gravel workings at Shotford [249 818] (just west of the district), where gravels have been dredged at up to 6 to 7 m below the level of the floodplain. The gravels at Shotford [246 813] have yielded a cold climate fauna of *Coelodonta antiquitatis* (woolly rhinoceros) and *Mammuthus primigenius* (mammoth) (Coxon, 1979; Coxon *in* Allen, 1984).

A sample from the gravels at Broadwash Farm road cutting [273 856] immediately west of the district gave a pebble count of flint 64%, quartz 27%, quartzite 4%, chalk 0% and others 4% (Coxon *in* Allen, 1984)

In 1967, excavations [2752 8437] for a sewer trench at Wortwell, near Harleston, revealed Ipswichian interglacial deposits lying non-sequentially below first terrace deposits and Head.

At Earlsham a temporary section [327 899] revealed first terrace deposits overlying peat; the peat has given a radiocarbon date of 11 210 years BP (Auton et al., 1985) and has yielded a beetle fauna indicative of cold climatic conditions (Taylor and Coope, 1985). Other late Devensian organic remains have been proved in hollows resting on the first terrace deposits in adjoining districts at Lopham Little Fen [042 794] (Tallantire, 1953), and in the Little Ouse valley to the west of Lopham (Coxon, 1978; Bradshaw et al., 1981). There is thus good evidence to indicate that the first terrace deposits are late Devensian in age.

River Blyth

Terrace deposits are restricted to a short stretch of valley between Holton [40 77] and Wenhaston [42 75] where small outcrops of three terraces have been recognised along both sides of the present valley. Post-depositional erosion has affected the terraces and their morphological features are poorly preserved. The lowest of the three terraces is contiguous with the alluvial silts and clays or the fringing peat where present. Its surface rises from about 2 m to over 4 m above OD away from the river. It is probable that the deposits extend beneath the alluvial deposits of the valley floor, although there are no borehole data to prove this. The deposits of the second terrace are separated from those of the first terrace by a small rise where Head has been mapped. The surface of the second terrace varies between about 4 and 6 m above OD. The surface of the third terrace ranges from about 4 m up to over 7 m above OD. Its outcrop is separated from that of the second terrace by slopes mantled in Head. All three terraces are composed of sand and gravel, the latter predominantly subangular to subrounded flint with smaller amounts of quartz and quartzite. No sections were visible at the time of survey.

There is no direct evidence as to the ages of the three terraces along the River Blyth, but by analogy with those of the River Waveney, to the north, they are likely to be ?late Anglian to Devensian.

River Deben

Two small deposits [305 566 and 310 564] have been mapped in the extreme south of the district about 7 km south-south-east of Framlingham. In isolation it is not clear how these relate to other terrace deposits within the Deben catchment, so they have not been assigned specific terrace numbers. Surface expressions indicate that the deposits comprise sand and gravel, the gravel component being predominantly flint with subordinate quartz and quartzite.

The terrace deposits are contiguous with the alluvial deposits of the River Deben and rise from about 10 m above OD along the floodplain margin to 2 or 3 m higher away from the river. Although there is no borehole evidence it is probable that the terrace sands and gravels extend beneath the alluvial deposits.

River Alde

Terrace deposits within the Alde catchment are restricted to a small area [330 576] about 6 km south-east of Framlingham within the valley of a tributary.

YARE VALLEY FORMATION

Nomenclature and distribution

This formation was defined by Arthurton et al. (1994) in the Great Yarmouth district. It occupies the floor of a buried valley system eroded into the underlying Solid formations (mostly Crag) and underlies the Breydon Formation (in most instances the Basal Peat). It is proved in boreholes only, and is restricted to the valleys occupied by Broadland rivers.

Evidence for the presence of the formation in the lower reaches of the Waveney valley (within the Great Yarmouth district) is fairly strong, but within the present district its existence is less certain. The presence of gravels beneath 'alluvial' sediments is generally recognised; Coxon (1979, p.20), for example, referred to 'the sands and gravels of the Waveney Floodplain and the associated alluvial sediments that cover these', and elsewhere (p.186) described the Waveney floodplain as '....a valley fill of sand and gravel overlain by...organic mud and clay'. Alderton (1983, pp.74–75) made reference to basal gravelly deposits in descriptions of transects across marshland from the upper Waveney down-valley towards the lower Yare. However, upstream of Burgh Marshes (gridline Northing 95) the Waveney valley contains many other gravelly deposits, exposed and buried, including the Aldeby Sands and Gravels, the Haddiscoe Sands and

Gravels, river terrace deposits and possibly Crag. Given the paucity of borehole data and absence of analytical information, it is impossible to distinguish these from the Yare Valley formation, and it is thus unreasonable to extrapolate the upstream limits of the Yare Valley Formation beyond Burgh Marshes.

The deposits comprise mainly flint gravel ranging from fine to coarse, with variable amounts of fine- to coarse-grained sand. In the type section, Runham/Yare borehole 8 (TG 50NW/480), the formation is 5.2 m thick and rests on Crag at 24.0 m below OD (Arthurton et al., 1994). In the district the formation is estimated at up to 2 m thick.

There is no direct evidence for the age of the Yare Valley Formation. Coxon (1979, p.214) referred to these deposits as Devensian, whilst Cox et al. (1989) suggested that deposition may have begun in the late Devensian. It is likely that at least some of the formation consists of fluvial sediments of late Devensian and early Flandrian age, deposited by rivers flowing within the now buried valley system which then drained central parts of East Anglia to the Southern North Sea Basin. The maximum age for the formation is more speculative. Funnell (unpublished, 1990) has argued that the general characteristics of the deposits imply a late Anglian age, but Arthurton et al. (1994) have maintained that the formation demonstrably postdates much of the Anglian succession, while admitting that glaciofluvial deposits of late Anglian age might be included.

If, as mapping in the district suggests, these 'buried' gravels merge westwards and upslope into exposed gravels such as the Aldeby, Haddiscoe and river terrace deposits, they would certainly range from late Anglian to Devensian, and possibly to Holocene, in age.

HEAD

Head deposits are produced by the mass movement of material downslope under periglacial conditions. During cold climatic conditions the downslope transport of materials is accelerated, due to reduced vegetation cover, increased precipitation and runoff, and cryoturbation. The deposits tend to accumulate along the lower slopes of valleys and in the floors of minor valleys where alluvial deposits are absent. Other processes such as hillwash and soil creep, which are active at the present time, also result in the movement of material downslope. Because it is impossible to distinguish the products of these later processes from deposits of periglacial origin they are all classified as Head on the published 1:50 000 maps. Much of the Head within the present district probably accumulated during the Devensian cold period.

The lithology of the Head deposits, because they are locally derived, tends to closely reflect that of their upslope source. For example, Head derived from the Crag will be predominantly sandy whereas that derived from the chalky till will be clay-rich, although the latter type tends to be decalcified and devoid of any chalk fragments.

Accumulations of Head may be up to several metres thick, with maximum thicknesses usually along valley floors. They are not shown on the maps where they are less than 1 metre thick.

In some localities the Head deposits are particularly gravelly; this is usually where they are derived from river terrace deposits or glaciofluvial gravels, but in some instances the source of the gravel is not obvious. Such occurrences are shown on the map by the Head symbol prefixed by the letter 'g'. Isolated spreads of gravelly Head occur near the margins of the Lowestoft till plateau, around Saint Cross South Elmham [29 83], Saint Margaret South Elmham [32 84] and Withersdale [27 81]. They consist of poorly sorted angular flint gravels in a sand matrix and locally contain pockets of well-sorted sand. These spreads give rise to a mounded topography, in crudely linear forms parallel to the modern drainage. They are probably remnants of the earliest postglacial fluvial sediments. On a smaller scale, cryoturbation pockets of gravelly material occur near the margins of the till plateau. At two localities in the Withersdale Street area [281 812 and 273 807] blocks of iron-cemented gravelly sand (ferricrete) were found as brash, probably derived from gravelly Head.

The upper part of a disused gravel pit [2632 8092] about one kilometre west of the district near Mendham Priory exposed the following:

	Thickness m
Topsoil	0.4
HEAD	
Gravel, unsorted, structureless, mainly subrounded to subangular flint with scattered quartzite pebbles in a clayey sand matrix	1.7
Sandy clay	seen

A trench [TM 2729 8228] dug near Punt's Lane, Mendham, about 200 m west of the district penetrated the following sequence:

	Thickness m
Topsoil	0.3
HEAD	
Sand, gravelly and clayey, orange, brown and fawn mottled, compact. Contains involution lobes of clean yellow medium-grained sand. Flint cobbles in basal 0.6 m	1.2
LOWESTOFT TILL FORMATION	2.5

Gravelly head forms an apron extending down below the Crag outcrop on the north side of the Blyth valley [401 773 to 415 773] from Holton to Blyford. It comprises loose sandy gravel derived largely from the Crag gravels, but with an input of sand and gravel from the glaciofluvial deposits. It is believed to have formed by sheet floods during a periglacial climate, rather than by solifluction.

OLDER BLOWN SAND

Older blown sand is shown on the published 1:50 000 map (Lowestoft sheet) around Outney Common [327 906]. Although the area has been landscaped, partly for a golf course and partly after shallow gravel diggings, the undulating topography seems to relate to natural sand ridges and arcuate dune features. The topography of the underlying first and second river terraces has been masked by blown sand comprising fine- to medium-grained sands with rare flint pebbles. At two localities [327 906 and 328 906] the blown sand has enclosed shallow peat basins; it also occurs as patches on the third river terrace. The position of these blown sands some 20 km inland from the present-day coast suggests that they owe their existence to an earlier environment; most likely they are late Devensian in age and relate to cold polar winds emanating from the contemporary ice sheet.

FIVE

Quaternary (Holocene): postglacial and present-day deposits

During the Devensian glacial maximum, 18 000 BP, it is estimated that around 5 per cent of the global water budget was locked up in the form of ice (Lamb, 1972; 1977); sea level as a consequence fell to as much as 120 m below its present level (Cronin, 1983). With the progressive but irregular amelioration of climate during the ensuing Holocene Epoch (from about 11 000 BP), melting of the ice sheets resulted in worldwide (eustatic) rises in sea level. Offshore data (Eisma et al., 1981; Jelgersma, 1979) show that the earliest marine incursion into the southern North Sea occurred before 8700 BP. The Holocene (Flandrian) transgression first entered the region from the English Channel through the Straits of Dover and, around 8300 BP, the Dogger Bank was submerged as the southern and northern North Sea became connected. Global studies of sea-level change confirm that the rise during the Flandrian was episodic (Anderson and Thomas, 1991) and marked by two intervals of rapid rise (Fairbanks, 1989).

Sea level was also affected by changes in the level of land brought about by glacio-isostatic loading and unloading. During the Anglian and later Devensian glaciations, the weight of ice caused the crust to be depressed, particularly in Scotland. The subsequent crustal rebound recorded in northern Britain by raised beaches was less strong in the south, where the original depression was less, and was also masked in southern and south-eastern England by differential downwarping. These crustal movements, combined with the eustatic rise of sea level, produced an estimated maximum rate of sea level rise of around 13 mm per year (Devoy, 1980) for the Thames estuary. This figure can be compared with the 2 mm per year relative rise since 4000 BP estimated for the Thames estuary and Norfolk by Shennan (1989).

Thus, at the start of the Holocene Epoch, the present-day marshland and river valleys of the district had a significantly different topography, draining over a land surface which extended out across an area where the North Sea is now. Evidence of the former river courses has been found by shallow seismic profiling and vibrocore sampling in the offshore areas. These reveal remnants of early Holocene sediments, often lying within depressions interpreted as former valleys (see tidal flat deposits and Figure 25). At about this time the coast at Great Yarmouth lay an estimated 7 km east of its present position (Arthurton et al., 1994).

The original valley floors of rivers now draining to marshland and tidal flats lie well below the present-day surface; for example the pre-Holocene valley of the Waveney lies 12 m below OD under Pete's Marsh [495 930] (Figure 26) whilst that of the proto-Blyth lies nearly 14 m below OD near Southwold (Brew et al., 1992). Initially, the rivers would have had steeper gradients in the incised valleys, and discharged into open embayments. As relative sea level rose, the lower reaches at least would have been subject to tidal influence, but as drainage base level rose, flow rates moderated and there was a corresponding broadening of the floodplains. During this time reedswamp and fen vegetation developed, particularly in the lower to middle reaches, producing organic debris now preserved in the Lower Peat. Although this deposit almost certainly was once a widespread layer overlying Pleistocene and older sands and gravels, remnants of it are discontinuous, probably because it was eroded by waves and tides during the subsequent marine incursion.

Between 8700 to 4500 BP, local sea level rose by between 22 and 26 m (Coles and Funnell, 1981; George 1992), the change being marked by the intrusion of tidal influence much further inland up the valleys. The fen vegetation gradually gave way to estuarine communities while estuarine clays and silts, the diachronous Lower Clay, were deposited. Adjacent to valley sides, reed bed communities probably persisted throughout the marine incursion, sustained by freshwater springs and runoff from the high ground, but towards the valley centres this vegetation would have given way to salt marsh and intertidal mud flats. The Lower Clay thickens seawards and dies out landwards; based on work in Broadland*, Coles and Funnell (1981) suggest the inland limit of this clay is at around 6 m below OD.

Around 4500 to 5000 BP, when local sea levels had risen to about 7 m below OD, the rate of rise slowed and freshwater conditions came to dominate. In the upper sections of valleys, where little or no Lower Clay sediments were deposited, reed beds gave way to alder-dominated woodland (George, 1992) whereas in the lower reaches floodplain swamp became established and the extensive Middle Peat accumulated. Some of the environmental change at this time was the result of wetter climatic conditions (Godwin, 1978) and locally, in the case of Broadland, was due also to the growth of a sand barrier across the mouth of the estuary (Coles and Funnell, 1981). Studies of analogous successions in the German Bight indicate that peat accumulation may have replaced mud deposition in some estuaries simply in response to a slowing in the rate of sea-level rise, at between 6500 and 2600 years BP (Ludwig et al., 1981).

Towards the end of the Middle Peat episode, about 2200 BP, peat accumulation gave way once more to

---

* Broadland takes its name from a series of shallow lakes or 'broads' which have developed in former peat diggings. They are linked to the lower tidally influenced reaches of three major rivers, the Waveney, the Yare and the Bure.

Figure 25 Distribution of alluvium, tidal flat deposits and Breydon Formation in the district.

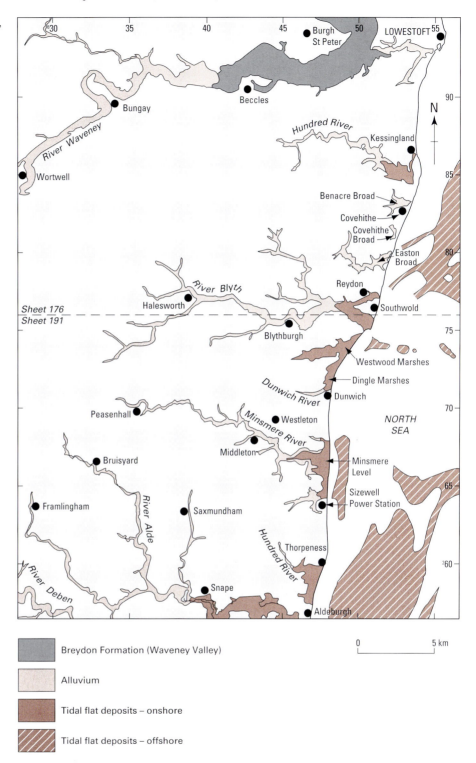

Breydon Formation (Waveney Valley)

Alluvium

Tidal flat deposits – onshore

Tidal flat deposits – offshore

0   5 km

deposition in marine-dominated mudflats and tidal channels, and the Upper Clay was laid down. The change of environment may have been due to several factors. Fairbanks (1989) has identified a second global increase in the rate of sea level change at around this time, but Arthurton et al. (1994) could find no local evidence for this. In Broadland (notably outside the district), the initial estuarine phase (muddy sediments) was overtaken by open-water conditions (shelly sands). This marine incursion may have been due to consolidation of soft deposits underneath (Middle Peat and Lower Clay; Arthurton et al., 1994), or to the disintegration of the protecting sand barrier (Coles and Funnell, 1981). In the Blyth estuary, Brew et al. (1992) recorded a gradual upward transition from freshwater peat, through saltmarsh, into high intertidal flats. They concluded that

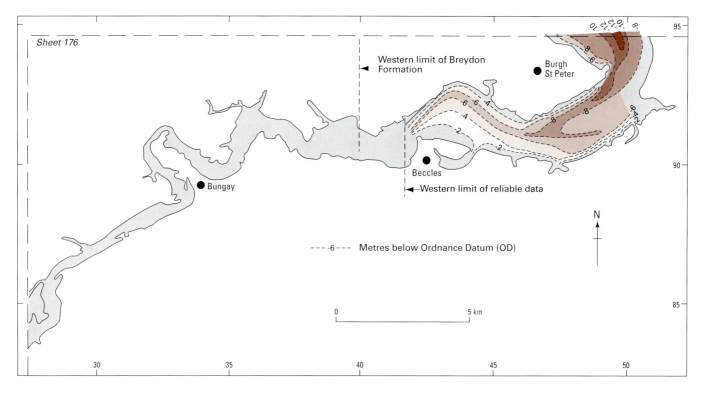

Figure 26 Contours (metres below OD) on the base of the Breydon Formation within the Waveney valley.

the lack of sand and the absence of fully marine fossils indicate a depositional environment some distance removed from open waters of the North Sea.

Whatever the causes, this second marine transgression penetrated deep into the hinterland, affecting tributaries as well as the main valleys; marine clay has been mapped as far as Beccles [425 905] in the Waveney valley and Brew (1990) recorded marine clay in the Blyth valley west of Blythburgh [452 755]. Historical records show that some of the estuaries were used as navigable waters probably before the first few centuries AD: the open estuary of Broadland (navigable at least to Burgh Castle [TG 475 045] in the Great Yarmouth district) was known to the Romans as *Gariensis Ostium*.

Where the mudflats were not continuously submerged, they were gradually colonised by salt-tolerant plants, and further up-valley the saltmarshes were succeeded by reed beds as tidal influence and salinities declined. Fringing fen or swamp vegetation thrived near to upland freshwater sources and locally, where groundwater seepage occurred well into the marshland, spring-fed islands of fen vegetation developed. In time, peat growth spread out and down-valley, coalescing to form the Upper Peat. The seaward extent of peat is unknown as the deposits suffered oxidation, weathering and wastage after human intervention (from about 1000 BP). However, it is likely that the Upper Peat was not established in the lowest reaches of the valleys and estuaries.

The onshore Holocene deposits of the district consist of alluvial sediments which become increasingly marine in nature eastwards towards the present-day shoreline (Figure 27). Thus freshwater clays, silts and peats pass down-valley into marshland comprising estuarine clays, silts, sands, and these give way to saltmarsh and tidal flats of the coastal fringe.

Offshore, modern marine sediments (muds, sands and pebbly sands) overlie former subaerial tidal mudflat deposits of early Holocene age, and Plio-Pleistocene Crag.

MARINE DEPOSITS

Marine deposits laid down in Holocene times are described under the three main environments evident at the present day.

Shoreface and beach deposits

The 41 km long coastline of the district is fringed by shoreface and beach deposits. On the landward side these may abut Quaternary cliffs or, where the land is low-lying, pass into blown sand or estuarine mud. The beach surface varies considerably. In some places thin veneers of fine- to medium-grained sand on gravel may be present intermittently throughout the year, in other places a top layer of clean shingle covers a sandy matrix present only a few centimetres below.

Beach sediments are naturally emphemeral, with periods of sediment accretion interrupted by storm events that result in a nett seaward removal of material. Typical shoreface profiles comprise shingle (mostly 10 to

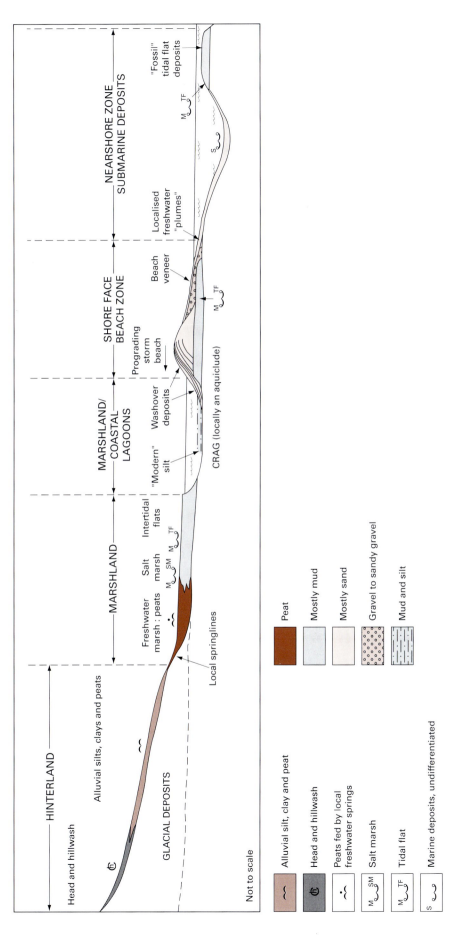

Figure 27   Schematic section showing the relationships of the Holocene deposits from the hinterland to the coast and the nearshore.

15 mm diameter) of well-rounded flint with subordinate quartz and quartzite, with a matrix of medium-grained sand. As noted above, gravel proportions are often highest in the upper parts of the vertical profile because interstitial sand is winnowed down by tidal and wave movements to produce sand-enriched basal layers. Excavations for the Sizewell 'B' nuclear power station exposed a section through beach deposits resting directly on the Red Crag. The basal contact dipped at 2 to 3° seawards but the deposits, consisting of interbedded sands and sandy gravels with well-rounded flint pebbles, were dipping in the same direction at 5 to 6°.

The shoreface and beach extends from the subtidal zone to around 2 to 3 m above OD, with higher levels on the landward side (up to about 7 m above OD) in coast-parallel storm ridges. The deposits form a wedge, trapezoid in cross-section and often resting on a wave-cut bench. The spring tidal range along the open coast varies from 1.9 m at Lowestoft to 2.1 m at Sizewell and the coast is therefore classified as micro- to mesotidal. This relatively small range, combined with the predominance of coarse material, produces narrow beaches with fairly steep intertidal gradients of 5 to 8°. Where the sand supply to the beach is more abundant, for example at Minsmere, the gradient may be reduced to around 4°.

Shoreface deposits mostly overlie Crag sands or Coralline Crag (see below), but adjacent to river valleys and low-lying marshland they overlie Holocene peats and clays, as near Minsmere Level [465 665], Corporation Marshes [495 735], near Benacre Broad [535828], near Beachfarm Marshes [530 845] and just south of Lowestoft Harbour [550 927].

The beach zone is typically 30 to 60 m wide, but may be less than 10 m where storm scours have stripped the beach to reveal bedrock, and is about 350 m wide along parts of Aldeburgh Bay [for example at 470 585]. The sediments are subject to an overall wave-induced southward migration or longshore drift. In historical times this has been responsible for the accumulation of the sediments forming the Lowestoft Denes spit, the several nesses (Lowestoft Ness, Benacre Ness and Thorpe Ness) down the coast, and the southwards deflection of the River Alde just outside the district.

The promontary of Thorpe Ness differs from the other two in that the underlying geology may control its location. The elongate outcrop of lithified Coralline Crag crosses the coastline at this point and is exposed on the sea bed just offshore. Studies of old maps show that the beach increased in width from 1883 to 1925, when it was 180 m wide. By 1978 it had decreased in width to 60 m at the Ness point (Robinson,1980) but by 1990 had increased again to over 120 m. By contrast to the long-term stability in the position of Thorpe Ness, Benacre Ness has migrated northwards by over 7 km since 1575, an average rate of around 19 m per year.

Present-day shore movements and sediment budgets have become important topics because of the damage that could be done by any rise of sea level that might result from global warming. Most of the beach material comes from the cliffs of Quaternary deposits in the district and to the north. Locally, rich supplies of pebbles are provided by the gravels of the Crag. Efforts to impede sediment movement include the use of regularly spaced groynes or concrete blocks and, most recently the dumping of massive Scandinavian metamorphic and igneous rocks along the stretch from the Coastguard Lookout [552 931] to North Beach [555 938].

Tidal flat deposits

Deposits of estuarine and marine silt and clay flank the lower reaches of rivers up to the limit of tidal flow. The principal areas underlain by these deposits are: Kessingland Level [530 850], east of Blythburgh [454 754] to Southwold's Town Marshes [501 754], Westwood Marshes [480 733], Minsmere Level [465 665], The Fens [458 596], and from Snape eastwards to Aldeburgh Marshes [455 567]. Drains and embankments have converted some of these once tidally submerged deposits into agricultural land, where sections in ditches or in river banks show well-laminated grey silt to mud. The deposits may be charged with plant debris, mainly roots.

Under stretches of open water, for example east of Blythburgh and Long Reach [420 570] and in the small 'broads', particularly Benacre Broad [530 830], these deposits are accreting at a very slow rate. Otherwise, tidal flat deposits are largely static, having accumulated during an earlier part of the Holocene when the coast lay farther to the east. A systematic programme of offshore seismic surveys and vibrocore sampling by BGS has confirmed that many of the major valleys which onshore are infilled by marine deposits continue eastwards into the offshore part of the district.

The largest of these valleys is the continuation of the Alde Valley, which lies to the south of the district. During the early Holocene when sea levels were lower the River Alde was not deflected southward by the modern Orford Spit and the drainage appears to have followed a more direct eastward or north-eastward path within the district. Seismic reflection profiles show that the valley here consists of a broad depression incised into the Crag sediments and up to 8 m deep. The original valley depth may have been greater but has been reduced by marine planation of interfluves during the Holocene transgression. The valley appears to be bounded to the north by the Coralline Crag outcrop which due to early lithification may have formed a positive feature during the early Holocene. Vibrocore sampling confirms that the valley is completely infilled with sediment comprising interbedded sands, silts and clays. Cores from just outside the district have recovered a basal freshwater peat overlain by silts and clays; the peat yielded occasional wood fragments of *Alnus*, and pollen which indicates ages between 6500 and 8000 BP. The overlying silts and clays contain molluscs including *Hydrobia*, *Cerastoderma edule*, *Littorina saxatilis* and *Ostrea edulis*, and abundant specimens of the foraminiferid *Ammonia beccarii*, indicative of a tidal flat environment. Further north the offshore extension of the Blyth valley is up to 8.5 m deep and also infilled with estuarine silts and clays but the sediments here include beds of sand and gravel.

Bank deposits

Linear sandbanks parallel to the coast are characteristic features of the inshore continental shelf around East Anglia. In the district several banks occur within a few kilometres of the coast.

The Sizewell–Dunwich Bank runs parallel to the coast and about 2 km offshore for a distance of about 11 km, from Thorpeness in the south to Southwold in the north. A central col separates the northern part (Dunwich Bank) from the southern part (Sizewell Bank). The bank is about 750 to 900 m wide and asymmetrical, with a landward face which slopes at approximately 1° and a gentler seaward slope of only 0.3°. It stands about 6 m high and is covered by only 4 m of water at low tide. The bank consists of fine- to very fine-grained well-sorted sand with some complete and comminuted shell material. Seismic profiles across it show the deposit to be around 9 m thick with internal landward dipping reflectors. Evidence from historic charts show that the bank has been migrating landwards at up to 10.7 m per year between 1867 and 1965 (Carr, 1979). It has also extended northwards at a rate of 49 m per year between 1824 and 1965. The bank sediments rest unconformably on the Crag and on early Holocene estuarine and tidal flat deposits which occupy shallow channels incised into the Crag. The bank is separated from the coast by a trough which reaches depths of over 9 m below low water, exposing early Holocene tidal flat deposits and Crag in places. This trough also separates the bank from the promontory of Thorpeness at its southern end.

Five kilometres further offshore, and in the extreme south-east of the district, Aldeburgh Napes is about 9 km long, just over 1 km wide and rises over 15 m above the adjacent sea floor.

A complex of small banks lies just offshore from Lowestoft Harbour. The largest of these, Newcome Sand, is 4 km long by 1.5 km wide and up to 7 m high. The bank appears to be connected to the shoreface at Benacre Ness and extends northwards to a point just east of Lowestoft Harbour entrance.

HOLOCENE VALLEY DEPOSITS

The Breydon Formation was defined by Arthurton et al. (1994) to include Holocene freshwater and estuarine sediments infilling a buried valley system beneath the marshland of Broadland. The name is retained here to maintain continuity with the Great Yarmouth district to the north, but its application is restricted to the Waveney valley (Figure 25). Holocene valley deposits elsewhere in the district are represented on the map in terms of the present-day surface deposits, alluvium and tidal flat deposits because the underlying sequences are poorly known.

Waveney valley: Breydon Formation

Marshland deposits within the Waveney valley were originally classified by Blake (1890) simply as Alluvium and are shown as such on the Old Series one-inch geological map. Within the middle to lower reaches of the Waveney (and elsewhere in Broadland), Lambert and Jennings (1960) established the essential marshland stratigraphy, recognising two main argillaceous divisions, the Upper Clay and Lower Clay, separated by the Middle Peat. Another peat, underlying the Lower Clay and forming the base of the succession, was termed the Basal Peat, and a further one, locally capping the Upper Clay, the Upper Peat. Later studies by Coles and Funnell (1981) of central Broadland and Alderton (1983) of the Waveney valley confirmed this succession, although Alderton proposed a locally-based nomenclature. Alderton's classification and the terminology adopted in the present work are shown in Table 9; see also Figure 28.

Outside the district the Breydon Formation is up to 22 m thick (Arthurton et al., 1994) but within the Waveney floodplain, based on Alderton's auger traverses, the formation ranges from more than 10 m thick around Share Marsh [4966 9361] to less than 5 m upstream of Marsh Lane [4516 9146].

The buried pre-Holocene surface underlying Waveney marshland (Figure 26) cuts through Anglian strata (mostly the Lowestoft Till and Corton formations) and the pre-Anglian Bytham Sands and Gravels into the Crag. Based on the available data, the valley floor possibly may

Table 9 Classification of marshland sediments in the Waveney valley.

Alderton's Classification (1983)	Type site: Augerhole SC6 [TM 4404 9298]	Waveney Valley: Augerhole BD 8/2 [TM 4890 9258]	Present survey after Arthurton et al. (1994)	
Waveney Formation	Stanley's Carr Peat Bed		Upper Peat	Breydon Formation
	Breydon Bed	Breydon Bed	Upper Clay	
	Aldeby Peat Bed	Burgh Peat Bed	Middle Peat	
		Oulton Bed	Lower Clay	
		Barnby Peat Bed	Lower Peat	

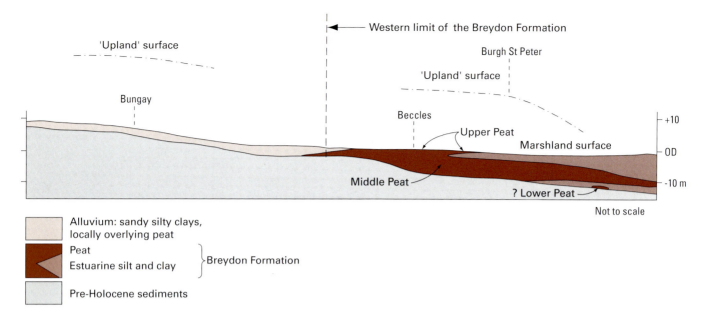

Figure 28 Schematic section through the upper Waveney valley showing the limits of the Breydon Formation.

take the form of an incised channel (locally lying at 7.2 m below OD) near Beccles. The incised shape becomes less pronounced downstream although the profile is noticeably asymmetrical, particularly down-valley towards Pete's Marsh where it lies deeper than 11 m below OD adjacent to the upland at Staithe [492 935].

Clay and silt

Clay and subordinate silt, formed in estuarine conditions, dominate the succession. These are the Lower and Upper Clays described above (Figure 28).

Within the district the **Lower Clay** is 0 to 2 m thick the unit thinning towards the buried valley margins (at about 6 m below OD) under Long Dam Level. Proved only in auger holes, it is a soft grey-black clay which becomes firm with depth.

The **Upper Clay** wedges out against Upper Peat near Beccles; it thickens eastwards to a maximum in the district of about 8 m and comprises the bulk of the Breydon Formation. Where exposed, the Upper Clay has a weathered upper layer, generally less than 1 m thick. This 'ripened soil' crust comprises silty to very silty clay, firm to very stiff, and pale grey in colour with distinctive tan mottling. It includes concretions of brown iron oxide and, commonly, traces of gypsum, plant fragments and rootlets. No other weathered layer is known at lower levels within the formation.

Beneath the weathered layer of the Upper Clay, two conspicuously different sedimentary facies are present. One is a mainly soft silty clay, pale to medium grey and rich in plant material which occurs as roots of the reed *Phragmites* in growth position, as scattered spongy fragments, or as comminuted debris imparting a bedding-parallel lamination to the sediments. This facies also includes sparse bivalve and gastropod shells. The

other facies (commonly underlying the first), comprises a soft to very soft (often liquid) silty clay, dark bluish or brownish grey to black in colour; it may be bioturbated and include interlaminated silts. The black colouration is due to finely disseminated iron pyrite (FeS_2) in diffuse layers, mottles and flecks. The pyrite formed diagenetically by the reduction of marine-derived dissolved sulphates by micro-organisms (including *Desulfovibrio desulfuvicans*) in anaerobic conditions (Price, 1980). Shells are locally common (comprising gastropods and thin-shelled bivalves) but disseminated plant material is rare.

'Soil ripening' appears to be at least in part a consequence of artificial drainage which causes irreversible shrinkage and the development of a strong, coarse prismatic soil structure, grading to blocky as ripening proceeds (Dent et al., 1976); it may be accompanied by the development of acid sulphate conditions.

Acid sulphate conditions are common in both peat and clay soils: the most severely affected soils are those around the edge of the marsh (Figure 29) where peat is adjacent to estuarine-derived clays (Burton and Hodgson, 1987). Commonly, acid sulphate horizons are characterised by a yellow mottling of jarosite ($KFe_3(SO_4)_2(OH)_6$) (Dent et al., 1976). The acidification results from the oxidation of pyrite in the sediment, which releases sulphuric acid and lowers the pH of the surrounding soil (Price, 1980). Generally this is not a problem to agriculture if the water table is maintained fairly close to the surface, for example alongside grazing pastures (O'Riordan, 1980), since oxidation is thereby inhibited. Liming of the soil is carried out to counteract the acid sulphate conditions.

Where pyrite is accompanied by calcium carbonate (from shell inclusions in the clays), the sulphuric acid

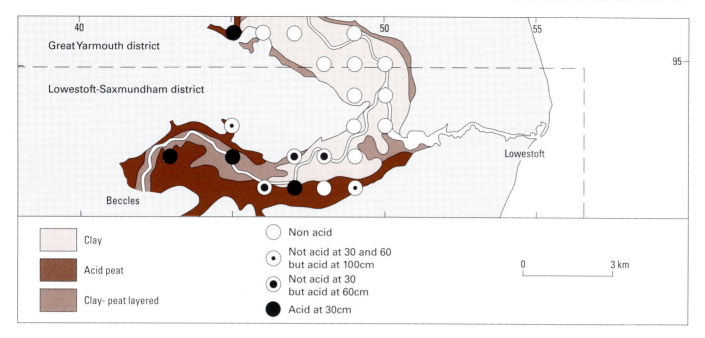

Figure 29 Waveney valley: distribution of soils liable to acidify if deep-drained. An acid soil is defined as one having an oxidised pH value of less than 4 or less than 3 after peroxide treatment. Diagram and detail from George, 1992. Source: Environmental Resource Management Ltd, and Trans Econ Ltd, 1981.

reacts to form gypsum (Price, 1980). Aeration of pyrite-rich sediments by drainage oxidises the pyrite to soluble iron and sulphuric acid. The process is enhanced by the bacterium *Thiobacillus ferroxidans*, and the soluble iron is precipitated as ferric hydroxide or ochre; in these instances ochre is observed in ditches and commonly is the cause of impeded drainage channels (Bloomfield, 1972; Dent, 1984, George, 1992).

The induced drainage of marshland has led to a progressive decrease in salinity of the underlying clay and silt as the salts have been leached out. Drainage of unstable sodium-rich clay soils results in deflocculation, soil movement and redeposition of clay minerals, blocking drains and clogging up fissures and pore spaces in the soils. Drainage failure and surface ponding can result (information from J Hazelden of the Soil Survey).

Soils developed on predominantly estuarine clay areas are classified by the Soil Survey as the Waveney, Newchurch and Wallasea series (Corbett and Tatler, 1970; Soil Survey of England and Wales, 1986; Burton and Hodgson, 1987; Hazelden, 1989).

Peat

The distribution of peat within the Breydon Formation is illustrated in Figure 28. In the lower reaches of the Waveney valley the peat outcrop mostly flanks the upland margin whilst west of Beccles Marshes [435 920] to Barsham Marshes [400 910] the outcrop extends across the whole valley. Upstream of Barsham Marshes the peat gives way to predominently alluvial sediments. The area of peat 'valley fill' coincides with an apparent break of slope

in the pre-Holocene valley profile (Figure 28). Although additional data are needed to confirm this feature, it appears that the slope break marks the approximate limit of marine incursions up the Waveney and is the site of preferential peat development. The bounding surfaces of the peat bodies are diachronous and their upper surfaces are erosional, at least in part. The stratigraphy of these units may be further complicated by consolidation after burial, particularly towards the centre of the valley.

Two main types of peat are recognised in the Breydon Formation: brushwood peat formed in freshwater, and *Phragmites* reed peat formed in brackish-water conditions. According to Burton and Hodgson (1987) the brushwood peat generally accumulated under eutrophic conditions and is commonly humified and woody, but where there is a transition to estuarine clay, sedges and reeds dominate producing a more fibrous peat which accumulated in fen and reedswamps associated with periods of high watertable.

The **Lower Peat** comprises a thin impersistent layer, entirely buried; within the Waveney valley, ^{14}C dates provided by Alderton (1983) for Lower Peat [at 4890 9258] are 6305 ± 55 BP (Q-2091) for the top of the unit at 8.6 m below OD, and 7750 ± 55 BP (Q-2092) at 9 m below OD. The impersistent nature of this layer may be a consequence of erosion following deposition. Alderton (1983) described the Lower Peat (her Barnby Peat) as always dry, compressed and humified so that the original composition is difficult to ascertain. At sites where analysis was possible, the peats comprised woody peat passing up into fen and reedswamp peat through to saltmarsh vegetation.

Like the Lower Peat, the **Middle Peat** is known only from auger sampling and boreholes. It is generally extensive, and is usually well defined between the Upper and Lower Clays. The upper surface often shows evidence of erosion. The Middle Peat is difficult to differentiate from the other two peats in the area where they coalesce (Figure 28). In the middle reaches of the Waveney below Beccles, the Middle Peat ranges from 3 to 5 m thick, whilst down-valley and out into the basin, the recorded thicknesses are less (in the range 2 to 3 m thick) probably due to dewatering and consolidation by increasingly thick cover of Upper Clay sediments.

Auger holes proving Middle Peat (the Burgh Peat Bed and Aldeby Peat of Alderton, 1983) show an upward succession through salt marsh, reedswamp and fen peats into an alder-*Salix* carr. A ^{14}C date for the base of this peat [at 4890 9258] is 4700 ± 55 BP (Q-2090), and one for the top of the unit is 2170 ± 55 BP (Q-2086) (Alderton, 1983).

The **Upper Peat** is mostly confined to discontinuous outcrops at the marshland fringe and is well developed where freshwater springs issue into embayments, for example near Wild Carr [443 908], or locally emanate into the marsh, as at Long Dam Level [460 915] and at Share Marsh [492 918]. The Upper Peat comprises mostly compact reed (*Phragmites*) and sedge (*Carex* and *Cladium*) peat, with some brushwood peat. Peat supplied with artesian water tends to be raised slightly above the marshland surface and is characterised by *Sphagnum*. A ^{14}C date for the base of the Upper Peat [at 4404 9298]) is 1755 ± 40 BP (Q-2183) (the Stanley's Carr Peat of Alderton, 1983).

Soils developed on well-established peat carrs of the marshland fringes are classified by the Soil Survey as the Adventurer's Series; where the peat thins or has mineral layers within it there are also the Prickwillow Series and the acid Mendham Series (Corbett and Tatler, 1970; Soil Survey of England and Wales, 1986; Burton and Hodgson, 1987; Hazelden, 1989).

Other valleys

Aside from the middle to lower reaches of the Waveney valley, where the Breydon Formation is distinguished, the district has over 100 km of river valleys and streams floored by alluvial sediments. These drainage systems are mostly freshwater, but become brackish towards the tidal reaches. Many valleys include both freshwater and estuarine sediments, with a stratigraphy comparable to that of the Breydon Formation; however, with the possible exception of the lower River Blyth catchment (described in detail by Brew, 1990; Brew et al., 1992), there is not enough subsurface information to enable the formation to be recognised.

The valleys are characteristically outsized in relation to the modern runoff; the interpretation here is that the valley profiles (in the hinterland at least) were initially determined during former glacial periods when runoff was greater. Valley widening would have resulted from repeated translations of meander bends. The broad lower reaches of valleys, merging into estuaries, have reached their present form through increased deposition of sediments under the influence of relative rises of sea level.

For the most part the valleys are floored by glacigene deposits (for example, the Aldeby Sands and Gravels underlying the Hundred River) although in the Bungay area, from Oaklands Farm [3210 8755] downstream to beyond Barsham Marshes [397 909], more recent river gravels are well developed and protrude through Holocene valley deposits as spreads of first terrace gravels. Upstream and upslope, the alluvial sediments imperceptibly merge and interdigitate with Head.

The Holocene valley deposits generally comprise unconsolidated layers of sand, silt, clay and organic material overlying or interbedded with gravel; they are mainly derived from Pleistocene deposits. Thicknesses range from 0 to 3 m in the higher reaches to greater than 10 m towards the limit of tides. On the published maps only the surface deposit is shown, normally alluvium or peat.

Alluvium

In the upper reaches of valleys, the alluvium often comprises an upper 'clayey to loamy' unit (perhaps overbank deposits) overlying a basal peat. A weathered layer at the top of the upper unit is characterised by stiff, tenacious, sandy to silty clays with root trace, and colour mottling in shades of orange, buff and grey. Below this weathered zone the clays become soft to glutinous and olive-brown to pale or dark grey, with occasional organic-rich pockets. Locally, finely disseminated iron pyrites is present as discrete black specks or streaks. The basal unit mostly comprises compact fibrous peat with traces of bark, roots and seeds (for example, alder and oak). Peaty clay intercalations are common. The peaty sediments are commonly waterlogged and may exude gas if penetrated. The shells of freshwater molluscs, including gastropods and bivalves, are present in places.

Alluvial fan deposits

A small area of alluvial fan deposits has been mapped adjacent to Head deposits in three valleys [403 748, 432 732 and 437 712]. These deposits comprise sandy flint gravels.

Peat

Downstream, and towards the floodplain margins, the alluvium grades into soft humic clay with linear bodies of surface peat flanking the valleys. Examples occur in the upper Waveney valley [285 857] and in the Blyth valley [415 772]. At the seaward end, freshwater alluvium gives way to brackish marsh peat adjacent to tidal mud flats, as at Latymer Dam [510 860] in the Hundred River and the River Minsmere [410 680 to 445 665] (Figure 28). These peats are mostly forming below reed beds including *Phragmites*, the material still harvested locally for thatching roofs.

Shell marl

A single outcrop [383 914] of shell marl occurs about 0.5 km south-west of Geldeston. It comprises a thin, calcareous, silty, humic clay with numerous shells. The deposit probable accumulated in a small backwater of the River Waveney.

BLOWN SAND

Localised bodies of blown sand occur as linear dunes aligned parallel or subparallel to the coast, or as veneers draping sea cliffs.

The dune belts are generally developed on the landward side of the foreshore, often perched on old storm beach ridges; they are mapped around Thorpeness, adjacent to the outfall of the River Blyth, at the outfall of the Hundred River near Benacre Ness [539 848], south of Kessingland Beach [535 859] and extensively along the Lowestoft Denes [553 943]. Minor occurrences, too small to depict on the 1:50 000 maps are at The Warren [522 806] and capping the beach ridge between Sizewell Hall [475 620] and Coney Hill [478 673]. Blown sand drapes are a feature of the cliff-line from Lowestoft Cliffs [550 948] to the College [552 935].

Blown sand ranges in thickness from a few tens of centimetres, where it drapes cliff features, to over 7 m in the thicker parts of dune belts. The sand is generally of fine-grained quartz, typically yellow-buff in colour and derived locally from beaches; a terrigenous component may be present where surface-blow from adjacent inland fields introduces seasonal veneers of soil to be incorporated in the dunes.

Unless stabilised by vegetation or human intervention, dunes tend to be ephemeral features; evidence from old maps of Suffolk suggests that many of the dune belts are not more than about 100 years old. The dune belt parallel to the coast just inland from Benacre Ness is truncated obliquely by a more recent erosive shoreline south of The Denes [536 840]; this dune belt was once part of a complex large spit extending to a point east of Covehithe, the former Covehithe Ness.

Much of the dune belt area has been disturbed by human activities; for example, the extensive area of the Lowestoft spit, described by Blake (1890) as overlain by 'concentric ridges [of blown sand] in front of the old cliffs' is today virtually flat ground given over to industrial buildings, caravan sites and car parks. Benacre dunes were disturbed during World War II when slit trenches and gun emplacements were dug along considerable lengths, and parts of the dune belt from Sizewell House to Minsmere Cliffs [478 678] include large blocks of concrete remaining from former World War II anti-tank emplacements.

GROUND MODIFIED BY HUMAN ACTIVITY

Made ground

This category represents material deposited by human agency on the original land surface. It falls broadly into two types: that created for constructional purposes and that which results from the disposal of waste materials. The former occurs in a variety of situations from road and rail embankments to screening mounds. Typically it consists of inert earth materials — clay, sand or gravel — but may also locally include builders' rubble, ash, cinders, and domestic and industrial waste. The second type includes waste tips at sites licensed and non-licensed with contents that may range from inert earth materials to domestic rubbish and industrial wastes. The depiction of made ground on the published 1:50 000 map is selective; many smaller areas, including many road and rail embankments, have been omitted for reasons of clarity. A third type of made ground is the veneer of material that accumulates in areas where human activity has been established for many years. This is not normally shown on the maps, but its presence in built up areas should be assumed unless site investigation proves otherwise.

Worked ground

This category is used for areas where the original land surface has been dug away and not backfilled. It includes road and rail cuttings as well as mineral workings, although most of the former have been omitted from the published 1:50 000 scale maps so as not to obscure the underlying geology. More detail can be found on the component 1:10 000 maps. Most of the mineral workings have been for aggregate, sand, or on a more limited scale, brickclay.

Many pits have been dug on the outcrop of the Lowestoft till. Some of the larger pits were probably dug as a source of brickclay, or as a source of calcareous material for 'marling' the adjacent acid soils developed on the sands and gravels. Most of the smaller pits, sited next to hedges and usually only a few metres in diameter, appear to have been dug as watering holes for livestock. Many of these have been backfilled as hedgelines have been removed.

At Lowestoft the docks area [552 929], the Outer Harbour [551 927], the Inner Harbour [541 927], Lake Lothing [530 930] and Oulton Broad [515 926] have all been excavated, to produce an artificial continuous waterway some 5 km in length. At Oulton Broad the excavations were dug as a source of peat.

Worked ground and made ground

The worked ground and made ground ornament is used on the map to indicate areas which have been excavated and subsequently backfilled. Such areas in the present district are predominantly former mineral workings, but also include backfilled watering holes for livestock (see above) and other excavations. No distincton is made between partially and completely backfilled excavations. The fill may include inert, domestic, industrial and commercial refuse. At Wangford the former gravel workings [470 778] in the Crag were lined with an impermeable membrane prior to use as a landfill site.

Landscaped ground

This category is used to depict areas where the original surface has been extensively remodelled, but where it is impractical or impossible to delineate areas of cut or made ground. At Henham Hall [450 772] a large area of landscaped ground has been mapped, where the surface has been remodelled by the addition of a large amount of peat, clay, silt, sand and gravel brought in from the Oulton Broad area some 15 km to the north.

Disturbed ground

The disturbed ground ornament is used for areas where made ground and ill-defined surface excavations are complexly associated with one another. In the present district these areas include such diverse sites as sewage treatment works and the nuclear power station at Sizewell.

SIX

Structure

The district lies towards the northern margin of the London–Brabant Massif (Wills, 1978). To the east and north-east lies the Southern North Sea Basin, a major sedimentary basin active through Mesozoic and Palaeogene times, and to the north is the Eastern England Shelf, a succession of Mesozoic sediments extending to the Market Weighton Block of east Yorkshire. The London–Brabant Massif consists of low-grade regionally metamorphosed and folded Precambrian and Lower Palaeozoic rocks which form part of the concealed Caledonide fold belt of eastern England (see Chapter 2) and remained a positive structure through Upper Palaeozoic and Mesozoic time.

Regional interpretation of aeromagnetic data indicates a depth to magnetic basement within the district of about 3 km or greater (Allsop, 1984, fig. 7). The magnetic material is likely to lie in the magnetic crystalline basement and most of the overlying 3 to 4 km thickness is likely to be Lower Palaeozoic sedimentary rocks. Thirty kilometres to the north-west of the district in north Norfolk a major granitic body has been postulated from geophysical evidence; between Norwich and Saxthorpe it is estimated to be only some 3.5 km below OD (Chroston et al., 1987).

In early to mid-Devonian times, the Precambrian and Lower Palaeozoic rocks of the region were folded and cleaved by the Acadian phase of the Caledonian orogeny (Soper et al., 1987). A major unconformity separates the Caledonian rocks of the massif from the overlying Mesozoic strata. West-north-westerly lineaments, as determined from the gravity and magnetic anomaly maps, dominate the district (Figure 4) and are clearly part of the regional Caledonide pattern for eastern England. Their associated gravity anomalies define the margins of alternating highs and lows and vary in amplitude from a few mGals to subtle features which are apparent only as elongations in the contours in Figure 4, or as a 'grain' on shaded relief plots (for example Lee et al., 1991, fig. 4).

CARBONIFEROUS TO EARLY MESOZOIC TECTONISM

Rocks of Carboniferous age are absent from the present district but occur in the north-eastern part of the adjacent Great Yarmouth district, where they thin on the margin of the London–Brabant Massif. Earth movements during the late Carboniferous and early Permian, related to the Variscan orogeny, resulted in folding and faulting of the Carboniferous and older rocks. Erosion then removed much of the Carboniferous sequence in the present district. For further information on the nature of the resulting pre-Permian unconformity the reader is referred to Tubb et al.(1986).

Subsidence and sedimentation commenced in the Southern North Sea Basin during early Permian times following a change to an east–west tensional stress regime with crustal extension and rifting. Most of the subsidence that took place during Permian and Triassic times probably resulted from thermal relaxation effects (Arthurton et al., 1994, p.91) following this phase of rifting.

In the adjacent Great Yarmouth district, to the north, progressive south-westward overlap of the Mesozoic sediments against the massif can be demonstrated; the massif maintained its positive status until early Cretaceous times, whereas recurrent subsidence took place in the Southern North Sea Basin through most of Permian, Triassic and Jurassic times.

Late Jurassic and early Cretaceous times marked another period of uplift and erosion in East Anglia, this one resulting in the formation of the late-Cimmerian unconformity (Fyfe et al., 1981; Ziegler, 1981). The uplift was probably caused by an isostatic adjustment of the crust, following the major extensional phase in the North Sea Basin (Chadwick, 1985a). Any Jurassic and Triassic rocks banked against the massif in the north of the district are likely to have been eroded off during this period.

CRETACEOUS AND LATER TECTONIC EVENTS

A phase of regional subsidence, caused by thermal relaxation, followed in the early Cretaceous (Barremian to Albian Stages: Chadwick, 1985b) and resulted in complete overstep of the London–Brabant Massif by the Gault during the Albian (Owen, 1971; Rawson et al., 1978). Throughout the Lower Cretaceous the London–Brabant Massif continued to influence sedimentation, but after this the massif underwent slow relative regional subsidence (Arthurton et al., 1994) which, coupled with high global sea levels (Vail et al., 1977), resulted in the deposition of a thick sequence (between 300 and 400 m) of Chalk (Chapter 2) within the district.

After the Chalk had been deposited, compression, uplift and basin inversion occurred (Chadwick, 1985c), all of which movements can be related to Alpine compression farther south. A north-eastward tilting of the region, and subsequent erosion, resulted in removal of progressively older Chalk to the south and west before deposition of the overlying Ormesby Clay Formation (Chapter 2) during the early Palaeogene. Offshore this unconformity becomes very marked, for example in the Hewett Gas Field, to the north of the present district, where most of the Chalk was removed before deposition of Palaeogene strata (Arthurton et al., 1994).

Wihin the district a thin development of Lambeth Group sediments (Chapter 2: formerly known as Woolwich and Reading Beds) has been proved in the Heath Farm Borehole near Halesworth; corresponding strata appear to be absent in the adjacent Great Yarmouth district to the north, suggesting that uplift and erosion may have affected the region in late Palaeocene times. Supporting evidence for this comes from two boreholes: in the Ormesby Borehole (just north of the Great Yarmouth district) the uppermost part of the Ormesby Clay is absent beneath sediments of the Thames Group, yet it is preserved in the Hales Borehole some 23 km to the south-west.

Following the deposition of the succeeding Thames Group (Chapter 2), eastward tilting occurred prior to the marine transgression associated with the development of the overlying Coralline Crag and Crag Group (Chapter 3). The thickness of the Crag Group varies greatly; the deposits are particularly thick within the south-west trending Stradbroke Basin which impinges on the centre-west side of the district (see Figure 4). This basin, along with similar ones elsewhere in East Anglia, is considered by Bristow (1983, Fig. 1) to be fault-bounded, the faulting having been initiated during Miocene uplift. Mathers and Zalasiewicz (1988) do not accept the fault-bounded nature of the basins and argue that the steep-sided margins to the troughs were produced by sea-bed scour.

Geophysical data have provided additional evidence on the nature of the Sudbury–Bildeston Ridge, a concealed Chalk bedrock high separating Crag-infilled basins. The gravity anomaly map of the region (Cornwell, 1985, and Figure 4) shows north-north-east-trending linear features coincident with some of the basin margins, suggesting the existence of structures at depth. Detailed geophysical surveys in the Thorndon area, just west of the district, have provided strong evidence that a local basement high underlies the ridge (Cornwell et al., 1996). Transient electromagnetic (TEM) soundings, in particular, have demonstrated a horst-like feature with an elevation some 100 m above the surrounding basement. The most likely interpretation is that the feature is defined by north-north-east-trending faults extending from within the basement through the Chalk. On the geological evidence of Bristow (1983), the faults probably extend into the Crag, although they have not been detected there using geophysical methods. The regional

gravity data (Figure 4) suggest that the structures extend north-eastwards to the Harleston area, where the Sudbury–Bildeston Ridge anomaly becomes more complex, and is associated with three local gravity highs.

In terms of its pronounced gravity signature the Sudbury–Bildeston Ridge structure appears to be unusual in East Anglia, but it is possible that other gravity lineaments, less well defined by the existing survey coverage, could reflect comparable structures. One example, about 10 km to the east, seems likely to represent a parallel structure. It is well defined over a distance of about 6 to 7 km and can be continued to the north-north-east as a series of dislocations in the gravity anomaly pattern. The lineament coincides with the northward projection of the Pettaugh Fault (Bristow, 1983). In the same area a magnetic lineament marks the eastern end of a large anomaly which extends across much of East Anglia, and which could represent the downward extension of the Sudbury–Bildeston Ridge faults in the deep magnetic basement.

North-east-trending displacements of the Chalk/Crag surface near Framlingham [28 63] and Peasenhall [35 69], indicated by borehole logs, but not obviously associated with gravity features, are interpreted as minor faults (see Figure 4) with maximum throws of several metres. Similar structures are probably present elsewhere but have not been detected because of an insufficient density of borehole data.

Dips of up to about 10° towards the south-east, recorded at a number of localities in gravels of the Crag Group, are depositional rather than tectonic in origin, and result from sedimentation within a beach plain environment (Hey, 1967).

Following the deposition of the Crag, there was an eastward downwarping of the coastal Crag sequence towards the North Sea Basin (Ziegler and Louwerens, 1979) and a compensatory uplift of the inland areas (Mathers and Zalaziewicz, 1988). A slight westward rise in the elevation of the clays in the upper part of the Crag inland from Easton Bavents is supporting evidence for such tilting. The current rate of subsidence in the coastal areas of countries bordering the southern North Sea is estimated to be in the order of 3 mm per year (West, 1968). The extent to which the current subsidence rate in East Anglia is affected by isostatic readjustment that followed the retreat of the icesheets is unknown.

SEVEN

Economic geology

MINERAL DEPOSITS

Sand and gravel

The district contains extensive resources of sand and gravel within the Crag Group and drift deposits, although they are commonly obscured by a blanket of glacial deposits except in coastal areas and the main river valleys.

To overcome the problem of assessing concealed sand and gravel, the British Geological Survey carried out a series of special surveys that involved drilling and particle size analysis of areas likely to contain resources. One of these surveys covered an area of some 85 km² in the north-western part of the district. During the survey the resource was drilled systematically with boreholes sited at 1 km intervals, and samples for particle size analysis were taken at 1 m intervals down each borehole. A report (Auton et al., 1985) includes maps which show areas where worthwhile mineral resources are present under workable depths of overburden, and gives details of the experimental results. The area assessed in this way does not include all the deposits of potential value in the district, however. Sand and gravel resources known to exist in the district are described under three headings: gravels associated with the River Waveney, gravels within the Crag Group, and sands within the Crag Group.

Gravels associated with the River Waveney

Sand and gravel workings here exploit the extensive river terrace deposits and associated glacial deposits along the sides of the Waveney valley. Active operations (1994) included two pits on adjacent sites at Flixton [304 861], a pit at Earsham [317 899], and two pits on adjacent sites at Aldeby [460 926 and 464 925], near Beccles.

These workings are typically located in gravel deposits, up to 10 m thick in places, which contain large (0.3 m) nodular flints that have clearly not been transported far from their source in the Chalk. These are collected and sold to flint knappers. Resources in the western part of the valley are well described in the report by Auton et al. (1985). They consist of river terrace deposits and under-lying sediments that occupy a channel approximately coincident with the present valley. The total mineral resource may be up to 18 m thick with a mean of 9 m, and the gravel content varies between 29% and 59% with an average of 41%. The pebbles consist mainly of flint, with minor quartz and quartzite accompanied sometimes by small amounts of chalk. The resource in the Waveney valley has been well exploited, which is a good indication of its value to the industry.

Gravels within the Crag Group

Within the Crag Group there are some extensive beach deposits rich in gravel (see Figure 2), which are currently exploited within the district. These were deposited in pre-glacial Pleistocene times and are made up predominantly of very well-rounded flint pebbles which had probably been subjected to long periods of attrition and reworking since they were first eroded from the Chalk.

The Crag gravels exploited at Thorington Quarry [420 728] have the appearance of typical beach shingle. The thickness of the gravel-rich deposit can be up to 10 m, but is typically 6.5 m. Workings at Henham [452 792] in the same deposits, now temporarily idle, are reported to be exploiting a deposit containing 80% gravel (coarser than 5 mm) and 20 m deep. Nearby, at Wangford Pit [469 780] another large-scale operation produces a high proportion of coarse aggregate.

Sands within the Crag Group

Extensive sand deposits in the Crag Group have attracted relatively little commercial interest. In general an excess of sand (fine aggregate) is produced by operations in East Anglia, and the main economic interest, therefore, lies in exploiting those formations with the highest gravel content. However, sands in the Crag Group are currently exploited at Wenhaston Quarry [412 768] in the Blyth valley near Halesworth. The pit contains extensive drying facilities as the product was formerly used as a foundry sand, and although some dried material is still produced for special purposes, the output is now mainly sold for mortar.

Brick clay

The Quaternary Chillesford Clay and other clays in the Crag Group at one time supported an extensive brick-making industry in parts of Suffolk where coal could be easily imported. However, during the present century the industry has contracted, and within the district it is now represented by only two brickworks and one extraction site.

At Cove Bottom [495 798] near Wrentham, clay is extracted from the Crag for use in a small brickworks nearby. The bricks are hand made, air dried (Plate 9) and fired in a single batch kiln. The product is a distinctive red-coloured facing brick which is valued for its aesthetic appearance and used over a wide area.

At Aldeburgh [443 578] there is a small brickworks that produces machine and hand-made bricks. Chillesford Clay was formerly extracted from a site next to the brickworks, but similar material is now obtained from a pit at Chillesford [387 525], outside the district.

In former times brick clay was also dug locally from the Lowestoft till, usually from the basal part which commonly has a lower chalk content than the remainder of the formation.

Plate 9 Drying shed at Cove Bottom Brickworks, near Wrentham (GS 568).

Peat

At Oulton Broad, in the north of the district, peat from the Breydon Formation was extracted for fuel between the 12th and 15th centuries AD, probably mostly during the 13th and 14th centuries (Smith, 1960). The peat workings were sited in the marginal parts of the marshlands where the Basal, Middle and Upper Peat layers coalesce. It has been suggested (Lambert and Jennings, 1960) that the combustible quality of the peat improved with depth, so that deep excavations were encouraged despite the additional difficulty of controlling the ingress of water. Away from the valley sides the limits of working were probably determined by the increasing thickness of Upper Clay overlying the Middle Peat (Lambert and Jennings, 1960). The later flooding of the peat diggings, or turbaries, has produced the extensive areas of water known as the Broads. The sides of the excavations are usually steep, vertical, or stepped, and the floors are generally horizontal except where the workings abut the valley sides. Typically the workings were 3 to 4 m deep. An examination of different editions of the Ordnance Survey map reveals the growth of much recent peat within the confines of the old workings.

HYDROGEOLOGY AND WATER SUPPLY

The water resources of the district are regulated by the Anglian Region of the National Rivers Authority. The district lies within hydrometric areas 34 and 35 and drainage is predominantly in an eastward direction by means of the River Alde, Minsmere River, River Blyth,

Latymere Dam, River Waveney and their tributaries. Mean annual precipitation is lowest along the coast and south-west of Bungay in the Waveney valley. It increases from 550 mm at Southwold to over 600 mm in the lower Waveney valley and over the higher ground inland. Average annual evapotranspiration is 480 mm.

The water resources of the district were first described by Whitaker (1906, 1921), and later surveys of the groundwater of the region can be found in publications by Woodland (1946) and East Suffolk and Norfolk River Authority (1971). The information is summarised on maps published by the Institute of Geological Sciences (1976, 1981). Historically the region relied predominantly on groundwater for supplies, Lowestoft being the only town to obtain its water solely from surface sources (Whitaker, 1906). It used Fritton Decoy, to the north of this district, which is probably largely fed by springs from the sands and gravels of the Corton and Lowestoft Till formations.

At present the total volume of licensed abstraction from the district is 31.3 million cubic metres per year (Mm³/a), of which 21.6 Mm³/a (69%) is groundwater. Table 10 shows the annual licensed abstraction figures from both surface and groundwater, subdivided by aquifer and usage. There is one river intake for public supply at Shipmeadow on the River Waveney [385 907] which accounts for 78% of the surface water abstracted. This, with the abstraction from Fritton Decoy in the district to the north, still supplies Lowestoft from surface water. Most of the remaining surface water (20%) is used for spray irrigation. The main aquifer in the district is the Chalk, but east of the Palaeogene boundary the Crag Group and glaciofluvial sands and gravels become more

Table 10 Quantity of water (millions of cubic metres per year) licensed to be abstracted annually from the Lowestoft and Saxmundham district (data supplied by Anglian Region, National Rivers Authority in October1994).

	Agriculture excluding spray irrigation*	Spray irrigation	Industry	Public and private supply†	Sand and gravel washing‡	Total
Drift	0.209(77)	1.600 (49)	0.057 (4)	2.810 (6)	0.909 (10)	5.585 (146)
Crag	0.086 (39)	1.686 (38)	0.246 (3)	2.263 (7)	0.249 (2)	4.530 (89)
Chalk	0.252 (87)	0.296 (10)	0.262 (6)	10.243 (12)	0.472 (2)	11.498 (117)
Total groundwater	0.547 (203)	3.555 (97)	0.565 (13)	15.316 (25)	1.630 (14)	21.613 (352)
Surface water	–	1.967 (57)	–	7.500 (1)	0.203 (2)	9.670 (60)
Total	0.547 (203)	5.522 (154)	0.565 (13)	22.816 (26)	1.833 (16)	31.283 (412)

* includes domestic and fishery
† includes Crown Estate
‡ includes amenity and water transfers
Numbers in brackets refer to number of licences

important. Of the total volume of groundwater abstracted 53% comes from the Chalk, 21% from the Crag and 26% from the drift deposits. The water from the drift comes predominantly from the glaciofluvial sands and gravels, with some contributions from other deposits. Groundwater is used predominantly for water supply (71%), with agriculture (including spray irrigation) using 19%, and sand and gravel washing and industry the remainder.

The deepest borehole in the district is at Lowestoft [5383 9263]; it is 558.4 m deep and was cased to the base of the Lower Greensand at 496 m. Up to 6.8 litres per second (l/s) of poor quality water with high sodium, sulphate and chloride ion concentrations (see Table 11) were pumped from the Silurian basement rocks, probably with a contribution from the Lower Greensand. The potentiometric surface was 14.6 m above ground level. An earlier analysis from the same borehole, when it was 496 m deep and probably drawing water from a combination of Lower Greensand, Upper Greensand and Chalk, is also given in Table 11; the quality was similar.

Chalk aquifer and overlying Palaeogene deposits

The Chalk underlies the whole of the district, but is nowhere exposed at the surface. In the west it is covered by Crag Group and younger deposits, except for two small areas where it is overlain by drift deposits alone. In the east, low permeability Palaeogene deposits intervene, confining water in the underlying Chalk. Figure 30 shows the potentiometric surface for both the Chalk and the Crag Group. The potentiometric surface is the minimum level relative to OD to which the water would be expected to rise in a borehole. Groundwater gradients in the Chalk typically vary from 1 in 250 to 1 in 1000. The highest groundwater levels occur around Framlingham where the potentiometric surface is about 27 m above

OD. Around Halesworth, to the east of the Palaeogene boundary, overflowing artesian conditions have been recorded with the potentiometric surface up to 2 m above ground level. Elsewhere water levels are generally within the overlying deposits. The zero contour (indicating where the potentiometric surface goes below sea level) is well inland and follows the Palaeogene boundary as far south as Beccles. Here the natural outlets from the Chalk aquifer, close to the Palaeogene boundary, are at low elevations and land drainage has locally reduced the water levels, with a consequent reduction in the piezometric surface of the confined aquifer. Further south there is a poor correlation between the surface water catchments of the Blyth, Minsmere and Alde and the groundwater units, and between Wrentham and Thorpeness the position of the zero contour is not known exactly. Apart from minor abstractions, there is no discharge from the confined aquifer to the sea, thus there is little groundwater flow east of the Palaeogene boundary. Seasonal variations in water levels in the Chalk beneath till are low, generally less than 1 m.

Permeability

The Chalk has a high primary porosity, generally over 30% (for the Upper and Middle Chalk), but the small size of the pores and interconnecting pore-throats give low matrix permeabilities. Consequently, the hydraulic conductivity of the matrix makes little contribution to the transmissivity of the aquifer (Price, 1987). Also, because of the small size of the pores, most of the stored water is held in the pores by capillary and molecular forces, and very little can move under the influence of gravity. Flow therefore depends on secondary permeability provided by fissures, which are often enlarged by solution. The interconnection of these fissures gives the rock its high permeability. Woodland (1946) suggested that as an area evolved topographically, increased infiltration occurred in the valleys as the less-permeable deposits were eroded

Table 11 Chemical analyses of groundwater from the main aquifers within or near the district.

	Lowestoft* 5383 9263	Lowestoft* 5383 9263	Benhall‡ 3529 6153	Saxmundham‡ 3790 6314	Beccles* 4191 9019	Aldeburgh 4545 5688	Leiston* 4445 6328	Syleham‡ 2091 7835	Southwold‡ 4857 7741	Kirby Cane† 3769 9266	Broome‡ 3560 9168
Location / National Grid reference (all TM)											
Type of source	Borehole	Borehole	Borehole	Borehole	Borehole	Shaft with headings	Shaft	Borehole	Shaft and heading	Borehole	Borehole
Aquifer	Silurian basement and ? Lower Greensand	Chalk, Upper Greensand and Lower Greensand	Chalk	Chalk	Chalk	Coralline Crag	Crag Group	Crag Group	glaciofluvial sands and gravels and Crag Group	glaciofluvial sands and gravels	river gravel
Date of analysis	20/5/1912	10/1907	18/4/1961	5/8/1991	20/7/1965	23/6/1992	13/10/1959	12/3/1992	1/8/1990	5/10/1989	13/3/1991
pH	—	—	6.9	7.2	7.2	7.1	6.9	7.3	7.3	7.5	7.2
Electrical conductivity (μmhos/cm)	—	—	1290	855	—	784	—	793	642	830	790
Total dissolved solids (mg/l)	8820	8644	870	—	645	—	550	—	—	680	—
Total hardness ($CaCO_3$) (mg/l)	2246	—	445	454	340	345	420	412	312	—	395
Bicarbonate (HCO_3^-) (mg/l)	128	133	305	380	317	268	402	366	208	360	332
Sulphate (SO_4^{2-}) (mg/l)	2438	2220	155	—	75	66	57.5	—	—	113	—
Chloride (Cl^-) (mg/l)	3200	3270	230	77	48	60	42	0.09	61	76.5	55
Nitrate (NO_3-N) (mg/l)	nil	0.3	0.3	<0.02	nil	16.0	4.4	0.4	10.5	7.4	0.5
Fluoride (F^-) (mg/l)	—	—	0.45	0.29	0.2	0.11	nil	0.18	0.10	—	0.18
Calcium (Ca^{2+}) (mg/l)	639	654	142	—	122	126.4	150	—	—	156.5	—
Magnesium (Mg^{2+}) (mg/l)	157	149	22	—	8	7	11	—	—	13.9	—
Sodium (Na^+) (mg/l)	2229	2209	127	36.7	38	33.1	21	22.3	26.7	52	30.5
Potassium (K^+) (mg/l)	54	—	12	4.9	—	3.7	—	3.8	3.2	3.1	5.0
Iron (total) (mg/l)	0.7	1.8	2.0	0.55	7.2	<0.017	0.08	10.7	0.03	0.31	0.23
Manganese (total) (mg/l)	—	—	absent	0.29	—	0.009	—	0.15	0.02	—	0.10
Silica (SiO_2) (mg/l)	—	—	23	—	10	—	8.4	—	—	—	—

* National Well Record Collection
† Anglian Water Services Ltd
‡ Essex and Suffolk Water

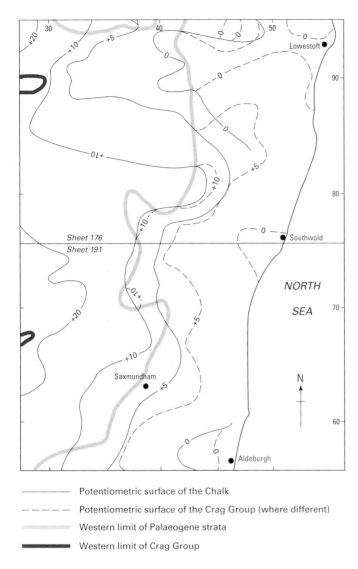

Potentiometric surface of the Chalk

---------- Potentiometric surface of the Crag Group (where different)

━━━━━ Western limit of Palaeogene strata

━━━━━ Western limit of Crag Group

Figure 30 Map showing contours in metres on the minimum potentiometric surfaces of the Chalk and Crag Group.

away, and that more solution then took place there than beneath the interfluves where recharge was limited. Thus higher transmissivities developed in the valleys and the process became self-perpetuating, with solution concentrating along some fissures at the expense of other less-permeable openings. It has been noted that the best-developed fissures occur in the zone of water table fluctuation, and within the top 20 m of the permanently saturated aquifer immediately below it. Price (1987) concluded, therefore, that the water table is constrained to stay within or near the zone of solution-enhanced fissures. However, other levels of fissures may have been developed during times when the water table was higher or lower.

Woodland (1946) correctly anticipated that the areas with largest yields would coincide with the Waveney, Blyth, Alde and Fromus valleys west of the Palaeogene clays. Here the rivers have cut through the till and hence

larger volumes of recharge have been able to reach the Chalk beneath glaciofluvial sands and Crag Group. Ineson (1962) suggested that valleys develop along lines of structural weakness and that subsequent unloading due to erosion produces local small-scale movements, with further shattering and increased permeabilities. The hard bands fracture more readily than the bulk of the Chalk and may often be the highest-yielding horizons; however, the fractures tend to close, and yields to decrease, with depth.

Where overlain directly by till, the top part of the Chalk, typically 10 to 12 m thick, may consist of brecciated Chalk in a fine Chalk matrix, giving a layer of low permeability; it is thought that this was produced by alternate freezing and thawing of the Chalk surface during the Pleistocene glacial episodes. Where saturated this layer is often soft and plastic, and difficult to drill through, but where it is above the water table it forms a hard layer.

Yields

The yield of a borehole is determined largely by the number and nature of the fissures it intersects. Hence most, but not all, of the highest-yielding boreholes are located in valleys. Yields of new boreholes can be increased significantly, and those of existing boreholes restored to their original values, by the addition of concentrated hydrochloric acid. This removes the low-permeability Chalk slurry from the walls of the borehole and from the fissures. Drawdowns are thereby decreased and pumping costs reduced. While Chalk will generally stand unsupported in a borehole, requiring only a few metres of lining to be installed near the surface, acid treatment is technically difficult unless the casing is continued below the water level (Mathers et al., 1993).

The highest-yielding borehole recorded in this district is at Barsham [4059 8951] where a borehole 91 m deep and 610 mm in diameter yielded up to 148 litres per second (l/s), from the Chalk beneath glaciofluvial sands and Crag Group, for a drawdown of 6.3 m during a 23 day test. Another high-yielding source is an acidised borehole at Shipmeadow [3880 9047] which yielded 138 l/s during an 8 hour test from a 70 m deep, 600 mm diameter borehole penetrating 47 m of Chalk beneath drift and Crag Group. The overlying deposits were lined out but probably contributed to this high yield by gravity drainage. A third borehole with a good yield is at Puddingmore, near Beccles [4191 9019]. This is 61 m deep and was tested at rates of between 80 and 107 l/s. However, its longer-term yield during an 8 day test was only 12.6 l/s for 32.5 m of drawdown. The 20 m of drift and Crag Group overlying the Chalk were again cased out. A borehole 686 mm in diameter and 61 m deep at Benhall [3537 6162], also with 20 m of lined-out glaciofluvial sands and gravels and Crag Group above the Chalk, yielded 69.5 l/s during an 8 day test for 26.2 m drawdown. Two 610 mm diameter boreholes at Walpole [3807 7567 and 3798 7566], both 70 m deep, yielded 44 l/s (for 14 days for a drawdown of 44.7 m) and 53 l/s (for 7 days for a drawdown of 21 m) respectively from the Chalk beneath 19 m of lined-out Crag Group. In the

Benhall and Walpole boreholes slotted casing was installed from the base of the lining tubes to the bottom of the borehole, and at Shipmeadow slotted casing was installed to a depth of 57 m.

In the interfluve areas groundwater is partially confined by till, and boreholes penetrating Chalk in these areas tend to have unsatisfactory yields, with water levels falling rapidly under test conditions; 100–150 mm diameter boreholes penetrating 30–40 m of Chalk beneath 20–65 m of overburden (till, glaciofluvial sands and gravels and Crag Group) commonly yield less than 1 l/s. These differences in yields are directly attributable to the differences in fissuring and transmissivity between valley and interfluve sites; transmissivity values in excess of 1000 m^2 per day have been measured at Barsham and Shipmeadow [3873 9043] while sites away from the valleys where the Chalk is overlain by till and Crag have values of less than 10 m^2 per day. The presence of Palaeogene deposits in the overburden does not seem to reduce the already small yields significantly; a 152 mm diameter trial borehole near Halesworth [3829 7771] yielded 13 l/s, for 4 days for a drawdown of 8.5 m, from Chalk beneath 15.2 m of Palaeogene strata and 19.8 m of younger deposits.

Quality

The quality of Chalk groundwater is generally good but hard (calcium bicarbonate type), with total hardness varying from 300 to 500 milligrammes per litre (mg/l) CaCO$_3$ and chloride ion concentrations of 50–100 mg/l over most of the area. Where overlain by Lowestoft till most of the hardness is caused by sulphates, and nitrate levels are generally not a problem. Beneath Crag Group the water is of similar quality, although sulphates may be lower. In the west of the area the total iron can exceed 4 mg/l and has reached as high as 7 mg/l because most of the recharge to the Chalk has come through the Crag and glaciofluvial sands and gravels. In the east of the district, beneath Palaeogene clays, sulphates increase and the chloride ion concentration exceeds 500 mg/l in an area 5 to 10 km wide parallel to the coast, and reached nearly 4000 mg/l in 1954 at Aldeby, near Beccles. The increase is rapid, going from 50 to 500 mg/l chloride over a distance of 1.5 km around Peasenhall. This is not caused by marine saline intrusion because there is no hydraulic connection between the Chalk aquifer and the sea, so must be due to connate water which has never been flushed out. Boreholes in a transitional zone contain shallow potable water underlain by denser saline water. Pumping in this area has to be carefully controlled, therefore, to ensure that the freshwater head is not reduced to the extent that saline water flows in from the east. There are several records of boreholes being partially backfilled in an attempt to reduce salinity. One is at Holton [3936 7723], where the bottom 7.5 m of a 56.5 m deep borehole, penetrating 21.5 m of Chalk beneath glaciofluvial sands and gravels and till, was cemented off within days of completion.

Lloyd et al. (1981) and Lloyd and Hiscock (1990) discussed the influence of the various drift deposits on the chemistry of Chalk waters. Lloyd et al. (1981) working in Essex, and Bath et al. (1985) in the present district and the area to the west, used tritium and carbon-14 analyses to date various Chalk groundwaters. At sites remote from river valleys, overlain by thick boulder clay, carbon-14 and tritium concentrations in the Chalk are low, indicating minimal modern recharge and ages of 10 000 to 30 000 years. In or near the river valleys, beneath Pleistocene sands, waters are more modern, reflecting the fact that the aquifer is directly recharged by infiltration through the sands. Therefore, although the potentiometric surface indicates that the hydraulic gradient is from the till-covered interfluves and towards the valleys, recharge actually takes place in the valleys. Song and Atkinson (1985) looked at the iron distribution in the Chalk west of the Crag Group outcrop, in the Bure catchment to the north of this district. Beneath till, iron concentrations are high due to the presence of reducing conditions, while in the valleys, beneath glaciofluvial sands, iron concentrations are lower and nitrates higher. Iron has often been precipitated out where waters from the two different groundwater systems have mixed (Lloyd and Hiscock, 1990).

Parker et al. (1987) showed that at Mattishall and Rushall, to the north-west and west respectively of this district, the Chalk waters from beneath till and glaciofluvial sands and gravels contained thermonuclear tritium but negligible nitrate, indicating that active recharge to the aquifer and in-situ bacterial denitrification are both taking place. At Mattishall there is clear physical, chemical and biological evidence for bacteriological denitrification within the groundwater system, causing nitrate to be removed as recharge is induced through the drift by pumping the Chalk. At Rushall, denitrification is thought to occur within the heavy till soils, as nitrate is absent even from the sandy shallow drift deposits.

Groundwater beneath the interfluves is therefore relatively old, free of nitrate, and in a reduced state often with high iron concentrations. By contrast, groundwater under or near valleys is modern, and high in nitrates and dissolved oxygen.

Palaeogene deposits

The low permeability Palaeogene deposits (Ormesby Clay Formation, Lambeth Group (formerly Woolwich and Reading Beds), Harwich Formation and London Clay Formation) are not used for water supplies and are normally cased out in boreholes that continue down into the Chalk. Their main hydrogeological significance is as an aquiclude confining groundwater in the underlying Chalk and acting as an impermeable base to the Crag Group and glaciofluvial sands and gravels aquifer above. Where these clays are present, groundwaters in the Chalk and Crag Group are hydraulically separate and have different potentiometric surfaces.

A 6.1 m deep shaft of 2134 mm diameter with two headings at Aldeburgh [4545 5688] yielded 18.8 l/s for 14 days for a drawdown of 0.2 m. The water has a total chloride ion concentration of 60 mg/l, and a nitrate concentration of 71 mg/l (Table 11) and is no longer used for public supply.

Crag Group aquifer and overlying deposits

The Crag Group (Norwich Crag and Red Crag combined) is considered to be a single water-bearing unit, although the Chillesford Clay and other argillaceous layers may produce perched water levels locally. The Coralline Crag is present only in the extreme south-east of the district. It forms an important local aquifer, but is not considered further in this account. The Crag Group forms a single aquifer with the immediately overlying glaciofluvial sands and gravels, and west of the Palaeogene boundary it is also in hydraulic continuity with the Chalk. In the interfluve areas groundwater is partially confined by the till cover. Where the Crag Group is hydraulically separated from the Chalk by Palaeogene deposits, Crag water levels are higher than the Chalk potentiometric surface and approximately reflect topography (Figure 30). Seasonal fluctuations are less than 1 m (East Suffolk and Norfolk River Authority, 1971) due to the high storage coefficient of the aquifer. The poorly consolidated nature of the Crag has limited its exploitation, and often where it directly overlies the Chalk it is lined out in boreholes. However, it is an important aquifer in the east where water from the Chalk is unsatisfactory in both quality and quantity. The Crag often comprises fine-grained uncemented sands which cause problems for well drilling and development. Even properly constructed boreholes can have relatively small yields due to the low permeability of the deposits. Higher yields are obtained from the coarser deposits. Boreholes must be properly designed and constructed to ensure that the slow ingress of fine-grained sand does not cause them to silt up after a short period of time.

Originally the Crag Group was exploited via shallow wells, commonly with associated adits. At Beccles [4174 8942], a well that was eventually continued down into the Chalk yielded (in 1890) 1.1 l/s from 3.7 m of Crag sands beneath 23.5 m of till, glaciofluvial sands and gravels and Crag clay. Generally this type of well is used for agricultural and domestic supplies and is pumped at low rates (often less than 1 l/s) so that sand entry is not a problem. Clarke and Phillips (1984) concluded that the best method of drilling in the Crag Group is to use reverse circulation, which limits the quantity of sand removed, and limits the area of resulting instability around the well, especially with percussion drilling. Sieve analysis of the material drilled is required to ensure that the gravel pack and screen are both of the appropriate size for the aquifer. Pre-formed sand and gravel packs bonded to the lining tubes are now available, and can be installed with or without conventional packs. The East Anglian Water Company, using a loose pack plus a pre-formed Hagusta pack and screen, found that virtually no material was pumped out of the aquifer, even during development with air-lift pumps at up to twice the production rate, and there was no need to top up the pack. A 61 m deep borehole at Shipmeadow [3854 8985], of 457 mm completed diameter, yielded 69.5 l/s from the Crag Group and the Chalk. Conventional slotted steel lining was used in the Chalk, while a pre-formed pack was used throughout the Crag. The whole of the borehole was surrounded by conventional gravel pack to prevent downward running sand entering the borehole via the Chalk section. Other sites, all obtaining high yields from the Crag beneath glaciofluvial sands and gravels, are at Lowestoft [5236 9422] where a borehole of 254 mm diameter and 79.6 m depth yielded 31.6 l/s for 13 days for a drawdown of 4.1 m; at Kirby Cane [3770 9258] where a 610 mm diameter borehole 13 m deep yielded 14.5 l/s for 1.2 days for a drawdown of 9.9 m; at Coldfair Green [4375 6082] where a 48.8 m deep borehole of 406 mm diameter yielded 56 l/s for 1.5 days for a drawdown of 9 m; and at Leiston [4414 6176] where a borehole 380 mm in diameter and 48.8 m deep yielded 51.5 l/s for 13.8 days for a drawdown of 11.6 m. Yields are commonly less than 10 l/s for 300 mm diameter boreholes and less than 5 l/s for 150 mm diameter boreholes.

Alternative abstraction systems have been used successfully in the Crag. They include well-point systems, which are used predominantly for spray irrigation and consist of a closely spaced series of shallow, narrow-diameter perforated tubes connected by a suction header to a single pump. They are restricted to areas where the water table is close to surface so that drawdowns do not exceed the atmospheric suction limit. The yield of an installation depends on the number of well-points, whose individual yields are likely to be around 0.7 l/s (East Suffolk and Norfolk River Authority, 1971). However, twelve 76 mm diameter well-points penetrating 6.1 m of Crag at Friston [439 583] yielded 15.2 l/s. Yields from the Crag Group are therefore a function of construction method as well as transmissivity; the latter ranging from less than 50 m² per day, to over 1000 m² per day around Leiston where the deposits are thicker.

At outcrop the carbonate hardness of Crag Group waters tends to be high, due to the decalcification of shell fragments, and where it is overlain by till the hardness tends to be very high (over 500 mg/l, of which 400 mg/l is non-carbonate). A feature of some Crag sources is high nitrate concentration, which is not necessarily of recent origin. Water from the Crag in a borehole that continued down into the Chalk at Southwold [5017 7614] had a nitrate concentration of 93 mg/l in 1888. Inland, the chloride concentration is generally less than 70 mg/l at outcrop and up to 150 mg/l beneath till. Near the coast the Crag is in hydraulic continuity with similar deposits offshore; the high chloride concentrations (several thousand mg/l) which occur in Crag wells on low ground near the coast, and in the vicinity of tidal rivers, are due to saline intrusion.

Crag waters are high in iron (up to 11 mg/l have been recorded) and manganese (up to 0.4 mg/l), and may also contain hydrogen sulphide. The removal of iron is a key supply issue and where treatment plants have been installed centrally for several wells, problems have been encountered with iron deposition in the untreated water pipelines (Clarke and Phillips, 1984). Small automated treatment plants are generally required at each wellhead. It is important that both the screen and pack are coated with acid-resistant material so that the whole borehole can be subjected periodically to rapid acid treatment with concentrated hydrochloric acid. Once the iron deposits have been removed, regular swabbing is needed to control the situation.

To the west of the district, water from the Crag Group at valley sites is modern in age, reflecting direct recharge through the overlying arenaceous Pleistocene deposits (Bath et al., 1985). Beneath till, away from the valleys, the water is older, but still less than 3000 years old and younger than the Chalk waters below till where the layer of low permeability cryoturbated chalk is a barrier to local infiltration.

Glaciofluvial sands and gravels provide small-scale aquifers. Generally they directly overlie the Crag Group and are in hydraulic continuity with it. They are highly permeable, allowing nearly all effective rainfall to percolate through to storage. Where they overlie the Crag they may locally be above the water table for part of the year, restricting their development as a source of supply. The resources have been exploited via shallow shafts and boreholes and some catchpits. The shafts are often supported by open brickwork or concrete liners, and boreholes require sand screens and filter packs.

At Wrentham [495 817], sited on glaciofluvial sands and gravels overlying clay within the Crag, 72 well-points, each 51 mm diameter and 10.7 m deep, yielded 20.8 l/s. Together the glaciofluvial sands and gravels and the Crag Group constitute a substantial, if complex, groundwater resource which has yielded large amounts of water. At Quay Lane [4857 7741] a 7.9 m well of 2743 mm diameter with a 15.2 m long heading yielded 10.1 l/s. At Alder Carr [4720 7717] a 5.9 m deep well of 1829 mm diameter with two 45.7 m long collecting pipes yielded 18.9 l/s for 0.7 m drawdown.

The presence of impersistent clay layers within the deposits promotes a layering of permeability and lateral movement of groundwater in the direction of valley discharge zones (Lloyd and Hiscock, 1990). The deposits have a high groundwater storage and in the west this is important in controlling recharge to the underlying Chalk aquifer. Further east, where vertical movement is impeded by Palaeogene clays, the water discharges into marshland fringes or side valleys.

Where the deposits are exposed at the surface, the water has a total hardness of 300 to 400 mg/l, with nitrate often from 5 to 10 mg/l of nitrogen. Water quality is affected by surface-derived contamination, which is often ferruginous, with levels up to 1.2 mg/l of total iron recorded. Beneath till, total hardness is more than 500 mg/l (mainly non-carbonate) and nitrate is generally negligible. Chlorides are generally less than 50 mg/l, except near the coast. The Pleistocene deposits also influence the quality of groundwater abstracted from the Chalk aquifer. Parker and James (1985) presented circumstantial evidence that recharge water to the Chalk is denitrified as it passes through the drift.

The till within the Lowestoft Till Formation is important hydrogeologically because it forms drapes of low-permeability material, often more than 25 m thick, over the underlying aquifers and thus protects them from the introduction of modern contaminants. However, the water levels in Chalk boreholes beneath till are locally higher than the Chalk rockhead, implying that some recharge must occur through the till. Supporting evidence for this is the presence of tritium in the Wensum valley (Parker and James, 1985). The existence of sand bodies within the till may permit a vertical component of recharge. However, Lloyd et al. (1981) concluded that in Essex, much of the water percolating through the till is diverted laterally through the Pleistocene sands to river valley outlets. A 10 m deep shaft of 1.2 m diameter into Lowestoft till at Ilketshall St Andrew [3867 8457] yielded 0.2 l/s on a 1 day test, from a sandy layer between 4.3 and 7.9 m depth. Drawdown was 1.2 m and the water good quality (chlorides 24 mg/l). Waters elsewhere are reported as being exceptionally hard (predominantly non-carbonate hardness). Yields are unlikely to be sustained due to the limited amount of recharge.

River gravels are of limited areal extent in this district and consequently are of little value as aquifers. At Broome Common [3559 9170] a 914 mm-diameter borehole through 7 m of gravels above 1.5 m of till yielded 15.2 l/s for 3 days for a drawdown of 4.2 m. At Outney Common, near Bungay [3264 9023] two 2133 mm diameter wells, 5.8 m deep and connected by a sump at 5.5 m, yielded 21.5 l/s from river gravels. Abstractions of this size are unusual. The quality of water from unpolluted sources is similar to, but harder than, that of any surface water with which they are in hydraulic continuity. However, the occurrence and shallow nature of these deposits mean that they are liable to pollution from the surface and from any water courses with which they are in hydraulic continuity.

Groundwater pollution comes from diffuse sources or point sources. Diffuse pollution is caused by the application of fertilisers and pesticides; the rise in nitrate concentrations in groundwaters is an example of this. This is less of a problem in this district than in other parts of East Anglia, due to the large expanses of till that overlie the aquifers. However, as described above, locally high nitrate concentrations are a problem in the Crag Group and more particularly in the glaciofluvial sands and gravels.

Point sources of pollution such as landfills and storage tanks represent more of a threat to groundwater quality. Historically, waste disposal has been carried out at a large number of small landfill sites in the district. Virtually all were located on till over Crag Group or on glaciofluvial sands and gravels over Crag. In 1994 there were four sites in the district accepting domestic wastes [524 883, 4710 7780, 467 927] or sewage [338 648]. The first of these is sited on Lowestoft till and is unlikely to present a hazard to groundwater; the others are sited on glaciofluvial sands and gravels overlying Crag sands and need to be closely monitored to ensure that leachate is not polluting the aquifer.

ENGINEERING GEOLOGY

No study of the engineering geology was undertaken during the survey of the district, but general information is available for some of the more problematical drift deposits. In particular, during the survey of the adjacent Great Yarmouth district, a study of the archived geotechnical data for all the Quaternary deposits mapped in the

Great Yarmouth urban area was carried out jointly by the British Geological Survey and the Geotechnical Consulting Group (civil engineering consultants) on behalf of the Anglian Water Authority. The information from this study relating to the alluvial deposits of the Breydon Formation has been summarized by Arthurton et al. (1994). The engineering properties of the till have been discussed in a number of papers including those of Boulton and Paul (1976), Denness (1974) and Little and Atkinson (1988).

With the exception of the Breydon Formation and other Holocene alluvial sequences the underlying soils do not pose many engineering problems, although differential subsidence may result where buildings span junctions between soil-types, or where there is variation within a soil. An example of the latter case would be where sand and gravel occurs within till. Throughout the district there are numerous small pits in the drift deposits, dug mainly for farm use; many are only a few metres across. In recent years many of them have been backfilled and can provide another source of potential differential subsidence if they remain undetected at the time of development. A watchful eye during the initial groundwork can often detect this problem in advance.

Engineering problems specific to the Breydon Formation and other alluvial sequences

The study in the Great Yarmouth district revealed a number of problems relating specifically to the Breydon Formation (Arthurton et al., 1994). Similar problems can be expected in alluvial sequences in the other river valleys within the district. The main problems are listed below.

a) base heave of shafts and trenches in soft clay
b) ingress of groundwater; control is by piping in shafts, trenches and tunnels
c) settlement due to movement of fines during dewatering
d) settlement or failure of embankments, flood defences and foundations
e) basement and near-surface pipeline flotation
f) problems with cast-in-place bored piles, namely shaft defects, necking, bulging in very soft clays and peats; base instability
g) problems with driven piles and pile groups in soft clays, namely over-driving, excess pore water pressures, heave, lateral displacements, negative skin friction
h) stability of shallow cuts
i) high-capacity trench support required
j) high sulphate and/or chloride content in groundwater near rivers and coast

The study also identified a number of problems likely to be encountered at the site investigation stage:

a) blowing in sands and gravels during standard penetration tests giving misleadingly low 'N' values
b) difficulties in obtaining good quality 'undisturbed' samples of peat and clay
c) difficulties in identifying thin layers of sand or clay
d) difficulties in locating the groundwater table and/or its fluctuation
e) collapse of trial pits

It should also be noted that the Upper Clay of the Breydon Formation has a very thin weathered crust, so that particular care needs to be exercised where the Upper Clay crops out at the surface.

COASTAL EROSION

The district has been noted for the loss of land along its seaward margin over a long period of time. Dunwich, now a small village, was once a large town, with many churches, and a harbour with a large fleet of warships. The cliffs at Covehithe are said to have once jutted out so far that they formed the easternmost point of England, a distinction now attributed to the town of Lowestoft. Plate 10 shows the devastation at Pakefield following a particularly fierce storm in 1936.

Quite detailed information on the rate of retreat of the coastline is available from the latter part of the nineteenth century up to the present. It is apparent from the data that the rate of erosion varies greatly from year to year, and also from place to place along the coast. Indeed, in some areas the coastline has been prograding eastwards for some considerable time.

Whitaker (1887) noted that the erosion is caused by the sliding down of masses of earth from the upper parts of the cliffs rather than by the undermining of the base of the cliffs by the sea. The cliff falls are caused by a variety of mechanisms that include freezing and thawing, drying and wetting, washing down of material by rain, and the blowing away of loose sand and gravel layers in the cliffs. The sea then removes the loose debris, enabling further falls to occur. Whitaker (1887) concluded that the best way to protect the coastline would be to encourage the accumulation of shingle, and to convert the steep cliffs into slopes with guarded bases. The taking of shingle and sand from the beaches, he considered, should be prohibited, as this could have a serious affect on the natural supply of protective shingle to other places farther along the coast. The observations and conclusions of Whitaker were somewhat ahead of their time. Only recently has it become generally accepted that the placing of artificial barriers to encourage shingle deposition at one locality may seriously affect the stability of the coastline elsewhere.

COVEHITHE   From 1878 to 1882 the average loss to the east of the Coastguard Station (now gone), was calculated to have been about 32 feet (9.8 m) per year (Whitaker, 1887).

Between 1882 and 1887 the cliff had further eroded back some 42 feet (12.9 m) in 5.25 years (allowing for some obliquity in measurement equivalent to about 2.4 m per year). Combining the two sets of figures the average loss over 9 years was about 5.5 m per year. Willson (1902) gives the cumulative loss at Covehithe in the 6 years prior to 1895 as 84 feet (25.6 m). In early

February 1993 there was evidence of many fresh falls along the cliff. Part of the field to the north of the lane which extends to the cliff had collapsed at some time after being sown with wheat in the previous autumn.

At Covehithe Broad the modern beach ridge is currently migrating westward over the Holocene peat forming the base of the broad, which at low tide is exposed on the seaward side of the shingle beach ridge resting on clays within the Crag Group. In February 1993 large isolated blocks of peat were present within the broad, apparently ripped up from the shore exposures and transported over the beach ridge during the high surge tide experienced during that month. The same tide was also responsible for breaching the beach ridge separating the broad from the sea. This resulted in a local lowering of the beach ridge and a partial drainage of the broad. A visit to the area on 23 February 1993 showed that at high tide the broad was inundated with sea water spilling through the breach.

EASTON BAVENTS   In 1841 Captain Alexander reported to the Geological Society that during the previous five years the retreat at Easton cliff averaged about 7 yards (6.4 m) per year. The coast at Easton Bavents, historically the most easterly land in England, has since retreated some two miles (3.3 km) westwards. In early February 1993 there were many fresh falls along this length of cliff but the high surge tide on 23 February removed much of the fallen material from the base of the cliffs.

At Easton Broad the modern shingle and sand beach ridge is migrating westwards over the Holocene peat forming the base of the broad. At low tide outcrops of the peat are exposed on the seaward side of the beach ridge.

SOUTHWOLD   At Southwold the cliffs have been well protected for many years. Whitaker (1887) mentions that no change could be noted while he was staying there. He quotes a Mr Redman as saying 'this place suffered much during the winter gale of 1862. Large portions of the cliff were then washed down, and this was no doubt induced by the constant degradation of the cliff south of Covehithe Ness'. On 16th May 1895 the gales, tides and rough sea cut away much land to create a new cove on the northern boundary of the town (Willson, 1902). Sole Bay, the site of a famous naval battle, remains as a name only on the Ordnance Survey maps, all evidence of any bay having disappeared into the sea.

BLYTH ESTUARY   Much of the estuary of the Blyth was drained, pumped and reclaimed for agriculture. However, in 1953 the tidal defences were breached, and much of the area has subsequently reverted to intertidal mud flats. Several public rights of way crossing the mud flats are depicted on Ordnance Survey maps, although the routes are now impossible to follow on foot.

DUNWICH   Willson (1902) states that by 1349 a huge slice of Dunwich, comprising upwards of four hundred houses, with many shops and windmills, were utterly devoured. By 1540, with the demolition of the parish of St Peter's, not a quarter of the old city of Dunwich was left standing. Between 1772 and 1880 the cliff appears to have retreated by 167 feet (50.9 m), giving a rate of about 18.5 inches (0.46 m) per year. In 1880, Whitaker (1887) records very little wastage, and mentions that the then current Ordnance Map and the previous Ordnance Map show very little difference, indicating that the rate of retreat just prior to 1880 was also not great.

Plate 10   Cliff erosion at Pakefield. An exceptionally high tide on November 30 1936 enabled the waves to attack the cliffs on the north side of Beach Street, causing strips of cliff to collapse and abandoned houses to fall into the sea (A 6943).

REFERENCES

Most of the references listed below are held in the Library of the British Geological Survey at Keyworth, Nottingham. Copies of the references can be purchased subject to current copyright legislation.

AGUIRE, E, and PASSINI, G. 1985. The Pliocene–Pleistocene boundary. *Episodes*, Vol. 8, No. 2, 116–120.

ALDERTON, A M. 1983. Flandrian vegetational history and sea-level change of the Waveney Valley. Unpublished PhD thesis, University of Cambridge.

ALLEN, P. 1984. *Field guide to the Gipping and Waveney valleys, Suffolk, May, 1984.* (Cambridge: Quaternary Research Association.)

ALLSOP, J M. 1984. Geophysical appraisal of a Carboniferous basin in northeast Norfolk, England. *Proceedings of the Geologists' Association*, Vol. 95, 175–180.

ALLSOP, J M. 1985. Geophysical investigations into the extent of the Devonian rocks beneath East Anglia. *Proceedings of the Geologists' Association*, Vol. 96, 371–379.

ALLSOP, J M. 1987. Patterns of late Caledonian intrusive activity in eastern and northern England from geophysics, radiometric dating and basement geology. *Proceedings of the Yorkshire Geological Society*, Vol. 46, 335–353.

ALLSOP, J M, and JONES, C M. 1981. A pre-Permian palaeogeological map of the East Midlands and East Anglia. *Transactions of the Leicester Literary and Philosophical Society*, No. 75, 28–33.

ANDERTON, R, BRIDGES, P H, LEEDER, M, and SELLWOOD, B W. 1979. *A dynamic stratigraphy of the British Isles.* (London: George Allen and Unwin.)

ANDERTON, J B, and THOMAS, M A. 1991. Marine ice-sheet decoupling as a mechanism for rapid, episodic sea-level change: the record of such events and their influence on sedimentation. *Sedimentary Geology*, Vol. 70, 87–104.

ARTHURTON, R S, BOOTH, S J, MORIGI, A N, ABBOTT, M A W, and WOOD, C J. 1994. Geology of the country around Great Yarmouth. *Memoir of the British Geological Survey*, Sheet 162 (England and Wales).

AUTON, C A, MORIGI, A N, and PRICE, D. 1985. The sand and gravel resources of the country around Harleston and Bungay, Norfolk and Suffolk. Description of 1:25 000 resource sheet comprising parts TM27, 28, 38, and 39. *Mineral Assessment Report of the British Geological Survey*, No. 145.

BADEN-POWELL, D F W. 1948. The chalky boulder clays of Norfolk and Suffolk. *Geological Magazine*, Vol. 85, 279–296.

BADEN-POWELL, D F W. 1950. Field meeting in the Lowestoft district. *Proceedings of the Geologists' Association*, Vol. 61, 191–197.

BAILEY, H W, GALE, A S, MORTIMORE, R N, SWIECICKI, A, and WOOD, C J. 1983. The Coniacian–Maastrichtian stages of the United Kingdom, with particular reference to southern England. *Newsletter of Stratigraphy*, Vol. 12, 29–42.

BALSON, P S. 1989. Tertiary phosphorites in the southern North Sea Basin: origin, evolution and stratigraphic correlation. 51–70 in *The Quaternary and Tertiary Geology of the Southern Bight, North Sea.* HENRIET, J P, and DE MOOR, G (editors). Ministry of Economic Affairs, Belgium Geological Survey.

BALSON, P S. 1990. The Neogene of East Anglia — a field excursion report. *Tertiary Research*, Vol. 11, 179–189.

BALSON, P S, and CAMERON, T D J. 1985. Quaternary mapping offshore East Anglia. *Modern Geology*, Vol. 9, 221–239.

BALSON, P S, MATHERS, S J, and ZALASIEWICZ, J A. 1993. The lithostratigraphy of the Coralline Crag (Pliocene) of Suffolk. *Proceedings of the Geologists' Association*, Vol. 104, 59–70.

BALSON, P S, and TAYLOR, P D. 1982. Palaeobiology and systematics of large cyclostome bryozoans from the Pliocene Coralline Crag of Suffolk. *Palaeontology*, Vol. 25, 529–554.

BANHAM, P H. 1970. Notes on Norfolk coastal sections. 2–4 in *Quaternary Research Association Field Guide, Norwich area.* BOULTON, G S (editor). (Norwich: Quaternary Research Association.)

BANHAM, P H. 1971. Pleistocene beds at Corton, Suffolk. *Geological Magazine*, Vol. 108, 281–285.

BANHAM, P H. 1975. Glacitectonic structures: a general discusssion with particular reference to the contorted drift of Norfolk. 69–94 *in* Ice ages: ancient and modern. WRIGHT, A E, and MOSELEY, F (editors). *Geological Journal Special Issue*, No. 6. (Liverpool: Seel House Press.)

BANHAM, P H. 1988. Polyphase glacitectonic deformation in the Contorted Drift of Norfolk. 27–32 in *Glaciotectonics: forms and processes.* CROOT, D G (editor). (Rotterdam: Balkema.)

BATEMAN, R M, and ROSE, J. 1994. Fine sand mineralogy of early and middle Pleistocene Bytham Sands and Gravels of midland England and East Anglia. *Proceedings of the Geologists' Association*, Vol. 105, 33–39.

BATH, A H, DOWNING, R A, and BARKER, J A. 1985. The age of groundwaters in the Chalk and Pleistocene sands of north-east Suffolk. *British Geological Survey Research Report*, WD/ST/85/1.

BATH, A H, and EDMUNDS, W. 1981. Identification of connate water in interstitial solution of Chalk sediment. *Geochimica et Cosmochimica Acta*, Vol. 45, 1449–1461.

BECK, R B, FUNNELL, B M, and LORD, A. 1972. Correlation of Lower Pleistocene Crag at depth in Suffolk. *Geological Magazine*, Vol. 109, 137–139.

BLAKE, J H. 1884a. Sections of Suffolk cliffs at Kessingland and Pakefield, and at Corton. Horizontal Section 128. Geological Survey of England and Wales.

BLAKE, J H. 1884b. Explanation of horizontal sections, Sheet 128. Sections of the Suffolk cliffs at Kessingland and Pakefield, and at Corton. Geological Survey of England and Wales.

BLAKE, J H. 1890. Geology of the country near Yarmouth and Lowestoft (explanation of Sheet 67). *Memoir of the Geological Survey (England and Wales)*.

BLOOMFIELD, C. 1972. The oxidation of iron sulphides in soils in relation to the formation of acid sulphate soils, and of ochre deposits in field drains. *Journal of Soil Science*, Vol. 23, 1–16.

BLUNDELL, D J. 1993. Deep structure of the Anglo–Brabant massif revealed by seismic profiling. *Geological Magazine,* Vol. 130, 563–567.

BOSWELL, P G H. 1914. On the occurrence of the North Sea Drift (Lower Glacial), and certain other brickearths in Suffolk. *Proceedings of the Geologists' Association,* Vol. 25, 121–153.

BOSWELL, P G H. 1916. The petrology of the North Sea Drift and Upper Glacial Brickearths in East Anglia. *Proceedings of the Geologists' Association,* Vol. 27, 79–98.

BOULTON, G S, and PAUL, M A. 1976. The influence of genetic processes on geotechnical properties of glacial tills. *Quarterly Journal of Engineering Geology,* Vol. 9, 159–194.

BOWEN, D Q, ROSE, J, McCABE, A M, and SUTHERLAND, D G. 1986. Correlation of Quaternary glaciations in England, Ireland, Scotland and Wales. *Quaternary Sciences Review,* Vol. 5, 299–340.

BRADSHAW, R H W, COXON, P, GREIG, J R A, and HALL, A R. 1981. New fossil evidence for the past cultivation and processing of hemp (*Cannabis sativa L.*) in eastern England. *New Phytologist,* Vol. 89, 503–510.

BREW, D S. 1990. Sedimentary environments and Holocene evolution of the Suffolk estuaries. Unpublished PhD thesis, University of East Anglia.

BREW, D S, FUNNELL, B M, and KREISER, A. 1992. Sedimentary environments and Holocene evolution of the lower Blyth estuary, Suffolk (England), and a comparison with other East Anglian coastal sequences. *Proceedings of the Geologists' Association,* Vol. 103, 57–74.

BRIDGE, D McC. 1993. Geological notes and local details for 1:10 000 sheets TM49NE, TM59NW and part of TM59NE (Somerleyton and Corton). *British Geological Survey Technical Report,* WA/93/92.

BRIDGE, D McC, and HOPSON, P M. 1985. Fine gravel, heavy mineral and grain-size analyses of Mid-Pleistocene glacial deposits in the lower Waveney Valley, East Anglia. *Modern Geology,* Vol. 9, 129–144.

BRIDGLAND, D R. 1988. The Pleistocene fluvial stratigraphy and palaeogeography of Essex. *Proceedings of the Geologists' Association,* Vol. 99, 291–314.

BRISTOW, C R. 1983. The stratigraphy and structure of the Crag of mid-Suffolk, England. *Proceedings of the Geologists' Association,* Vol. 94, 1–12.

BRISTOW, C R. 1990. Geology of the country around Bury St Edmunds. *Memoir of the British Geological Survey,* Sheet 189 (England and Wales).

BRISTOW, C R, and COX, F C. 1973. The Gipping Till: a reappraisal of East Anglian glacial stratigraphy. *Quarterly Journal of the Geological Society of London,* Vol. 129, 1–37.

BRITISH GEOLOGICAL SURVEY. 1985. *East Anglia. Sheet 52°N–00°. Solid Geology. 1:250 000 Series.* (Southampton: Ordnance Survey for the British Geological Survey.)

BROADS AUTHORITY. 1981. Acid sulphate soils in Broadland. *Broads Authority Research Series,* No. 3.

BROMLEY, R G, SCHILZ, M G, and PEAKE, N B. 1975. Paramoudras: giant flints, long burrows and the early diagenesis of chalks. *Det Kongelige Danske Videnskabernes Selskab,* Vol. 20, 1–31.

BULLARD, E C, GASKELL, T F, HARLAND, W B, and KERR-GRANT, C. 1940. Seismic investigations of the Palaeozoic floor of east England. *Philosophical Transactions of the Royal Society of London,* Series A, Vol. 239, 29–94.

BURTON, R G O, and HODGSON, J M (editors). 1987. Lowland peat in England and Wales. *Soil Survey Special Survey,* No.15. (Harpenden: Soil Survey of England and Wales.)

CAMERON, T D J, BONNY, A P, GREGORY, D M, and HARLAND, R. 1984. Lower Pleistocene dinoflagellate cyst, foraminiferal and pollen assemblages in four boreholes in the southern North Sea. *Geological Magazine,* Vol. 121, 85–97.

CAMERON, T D J, CROSBY, A, BALSON, P S, JEFFERY, D H, LOTT, G K, BULAT, J, and HARRISON, D J. 1992. *United Kingdom offshore regional report: the geology of the southern North Sea.* (London: HMSO for the British Geological Survey.)

CAMERON, T D J, STOKER, M S, and LONG, D. 1987. The history of Quaternary sedimentation in the UK sector of the North Sea Basin. *Journal of the Geological Society of London,* Vol. 144, 43–58.

CANDE, S C and KENT, D V. 1992. A new magnetic polarity time scale for the late Cretaceous and Cenozoic. *Journal of Geophysical Science,* Vol. 97, 13917–13951

CANDLER, C. 1889. Observations on some undescribed lacustrine deposits at Saint Cross, South Elmham in Suffolk *Quarterly Journal of the Geological Society of London,* Vol. 45, 504–510.

CARR, A P. 1967. The London Clay surface in part of Suffolk. *Geological Magazine,* Vol. 104, 574–584.

CARR, A P. 1979. Sizewell–Dunwich Banks field study. Topic Report 2. Long-term changes in the coastline and offshore banks. *Report of the Institute of Oceanographic Sciences,* No. 89. 25pp.

CARR, A P. 1981. Evidence for the sediment circulation along the coast of East Anglia. *Marine Geology,* Vol. 40, M9–M22.

CHADWICK, R A. 1985a. End Jurassic–early Cretaceous sedimentation and subsidence (late Portlandian to Barremian), and the late-Cimmerian unconformity. 52–56 in *Atlas of onshore sedimentary basins in England and Wales.* WHITTAKER, A (editor). (Keyworth: British Geological Survey, and Glasgow and London: Blackie.)

CHADWICK, R A. 1985b. Cretaceous sedimentation and subsidence (Albian-Aptian). 57–58 in *Atlas of onshore sedimentary basins in England and Wales.* WHITTAKER, A (editor). (Keyworth: British Geological Survey, and Glasgow and London: Blackie.)

CHADWICK, R A. 1985c. Cenozoic sedimentation and subsidence and tectonic inversion. 61–63 in *Atlas of onshore sedimentary basins in England and Wales.* WHITTAKER, A (editor). (Keyworth: British Geological Survey, and Glasgow and London: Blackie.)

CHARLESWORTH, E. 1835. Observations on the Crag Formation and its organic remains: with a view to establishing a division of the Tertiary strata overlying the London Clay in Suffolk. *Philosophical Magazine,* Vol. 7, 81–94.

CHRISTENSEN, W K. 1991 Belemnites from the Coniacian to Lower Campanian chalks of Norfolk and southern England. *Palaeontology,* Vol. 34, 695–749.

CHROSTON, P N. 1985. A seismic refraction line across Norfolk. *Geological Magazine,* Vol. 122, 397–401.

CHROSTON, P N, ALLSOP, J M, and CORNWELL, J D. 1987. New seismic refraction evidence of the origin of the Bouguer anomaly low near Hunstanton, Norfolk. *Proceedings of the Yorkshire Geological Society,* Vol. 46, 311–319.

CHROSTON, P N, and SOLA, M A. 1975. The sub-Mesozoic floor in Norfolk. *Bulletin of the Geological Society of Norfolk,* Vol. 27, 3–19.

CHROSTON, P N, and SOLA, M A. 1982. Deep boreholes, seismic refraction lines and the interpretation of gravity anomalies in Norfolk. *Journal of the Geological Society of London*, Vol. 139, 255–264.

CLARKE, K B, and PHILLIPS, J H. 1984. Experiences in the use of East Anglian sands and gravels ('Crags') as a source of water supply. *Journal of the Institution of Water Engineers & Scientists*, Vol. 38, 543–549.

CLARKE, M R. 1983. The sand and gravel resources of the country around Woolpit, Suffolk. Description of 1:25 000 sheet TL96. *Mineral Assessment Report of the Institute of Geological Sciences*, No. 127.

CLARKE, M R, and AUTON, C A. 1982. The Pleistocene depositional history of the Norfolk–Suffolk borderlands. *Report of the Institute of Geological Sciences*, No. 82/1, 23–29.

CLAYTON, C J. 1986. The chemical environment of flint formation in Upper Cretaceous chalks. 43–54 *in* The scientific study of flint and chert. *Proceedings of The Fourth International Flint Symposium.* SIEVEKING, G DE G, and HART, M B (editors). (Cambridge: Cambridge University Press.)

CLEEVELY, R J, and MORRIS, N J. 1987. Introduction to mollusca and bivalves. *In* Fossils of the Chalk. *Palaeontological Association Field Guides to Fossils No. 2.* Smith, A B (editor). (London: Palaeontological Association.)

COLES, B P L. 1977. The Holocene foraminifera and palaeogeography of Central Broadland. Unpublished PhD thesis, University of East Anglia.

COLES, B P L, and FUNNELL, B M. 1981. Holocene palaeoenvironments of Broadland, England. *International Association of Sedimentologists Special Publication*, No. 5, 123–131.

CORBETT, W M, and TATLER, W. 1970. Soils in Norfolk Sheet TM49 (Beccles North). *Survey Record No. 1. Soil Survey of England and Wales.*

CORNWELL, J D. 1985. Applications of geophysical methods to mapping unconsolidated sediments in East Anglia. *Modern Geology*, Vol. 9, 187–205.

CORNWELL, J D, KIMBELL, G S, and OGILVY, R D. 1996. Geophysical evidence for basement structure in Suffolk, East Anglia. *Journal of the Geological Society of London*, Vol. 153, 207–211.

CORNWELL, J D, and MCCANN, D M. 1991. The application of geophysical methods to the geological mapping of Quaternary sediments. 519–526 *in* Quaternary engineering geology. FORSTER, A, CULSHAW, M G, CRIPPS, A C, LITTLE, J, and MOON, C F (editors). *Geological Society Engineering Geology Special Publication* No. 7.

COSTA, L I, and MANUM, S B. 1988. The description of the interregional zonation of the Palaeogene (D1–D15) and the Miocene (D16–D20). 321–330 *in* The Northwest European Tertiary Basin. VINKEN, R (compiler). *Geologisches Jahrbuch*, Vol. A100.

COX, F C. 1985. The East Anglia Regional Geological Survey: an overview. *Modern Geology*, Vol. 9, 103–115.

COX, F C, GALLOIS, R W, and WOOD, C J. 1989. Geology of the country around Norwich. *Memoir of the British Geological Survey*, Sheet 161 (England and Wales).

COX, F C, HAILWOOD, E A, HARLAND, R, HUGHES, M J, JOHNSTON, N, and KNOX, R W O'B. 1985. Palaeocene sedimentation and stratigraphy in Norfolk, England. *Newsletters on Stratigraphy*, Vol. 14, 169–185.

COX, F C, and NICKLESS, E F P. 1972. Some aspects of the glacial history of central Norfolk. *Bulletin of the Geological Survey of Great Britain*, Vol. 42, 79–98.

COXON, P. 1979. Pleistocene environmental history in central East Anglia. Unpublished PhD thesis, University of Cambridge.

CRONIN, T M. 1983. Rapid sea-level and climatic changes: evidence from continental and island margins. *Quaternary Science Reviews*, Vol. 1, 177–214.

CURRY, D, ADAMS, C G, BOULTER, M C, DILLEY, F C, EAMES, F E, FUNNELLS, B M, and WELLS, M K. 1978. A correlation of Tertiary rocks in the British Isles. *Geological Society of London Special Report*, No. 12.

DALTON, W H, and WHITAKER, W. 1886. The geology of the country around Aldeborough, Orford, and Woodbridge. *Memoir of the Geological Survey.*

DENNESS, B. 1974. Engineering aspects of the chalky boulder clay at the new town of Milton Keynes in Buckinghamshire. *Quarterly Journal of Engineering Geology*, Vol. 7, 297–309.

DENT, D L. 1984. An introduction to acid sulphate soils and their occurrence in East Anglia. 35–51 *in* Soil acidification in SE England. BURNHAM, C P, and MOFFAT, A J. (editors). *Seesoil*, Vol.2.

DENT, D L, DOWNING, E J B, and ROGAAR, H. 1976. Changes in the structure of marsh soils, following drainage and arable cultivation. *Journal of Soil Science*, Vol. 27, 250–265.

DEVOY, R J N. 1980. Postglacial environmental change and Man in the Thames Estuary: synopsis. 37–55 *in* Archaeology and coastal change. THOMPSON, F A (editor). *The Society of Antiquaries, London, Occasional Paper (New Series)* No. 1.

DIXON, R G. 1972. A review of the Chillesford Beds. *Bulletin of the Ipswich Geological Group*, Vol. 11, 2–9.

DIXON, R G. 1978. Deposits marginal to the Red Crag basin. *Bulletin of the Geological Society of Norfolk*, Vol. 30, 92–104.

DIXON, R G. 1979. Sedimentary facies in the Red Crag (Lower Pleistocene), East Anglia. *Proceedings of the Geologists' Association*, Vol. 90, 117–132.

EAST SUFFOLK AND NORFOLK RIVER AUTHORITY. 1971. First survey of resources and demands, 2 volumes.

EISMA, D, MOOK, W G, and LABAN, C. 1981. An early Holocene tidal flat in the Southern Bight. 229–237 *in* Holocene marine sedimentation in the North Sea Basin. NIO, S-D, SHÜTTENHELM (sic), RTE, and WEERING, T J C E VAN (editors). *Special Publication of the International Association of Sedimentologists*, No. 5.

ELLISON, R A, KNOX, R W O'B, JOLLEY, D W, and KING, C. 1994. A revision of the lithostratigraphical classification of the early Palaeogene strata of the London Basin and East Anglia. *Proceedings of the Geologists' Association*, Vol. 105, 187–197.

ELLISON, R A, and LAKE, R D. 1986. Geology of the country around Braintree. *Memoir of the British Geological Survey*, Sheet 223 (England and Wales).

EVANS, C J, and ALLSOP, J M. 1987. Some geophysical aspects of the deep geology of eastern England. *Proceedings of the Yorkshire Geological Society*, Vol. 46, 321–333.

EYLES, N, EYLES, C H, and MCCABE, A M. 1989. Sedimentation in an ice-contact subaqueous setting: the mid Pleistocene 'North Sea Drift' of Norfolk, UK. *Quaternary Science Reviews*, Vol. 8, 57–74.

FAIRBANKS, R G. 1989. A 17 000-year glacio-eustatic sea-level record: influence of glacial melting rates on the Younger Dryas event and deep-ocean circulation. *Nature, London*, Vol. 342, 637–642.

FLETCHER, T P, and WOOD, C J. 1978. Chapter 15, Cretaceous rocks. *In* Geology of the Causeway Coast, Vol. 2. WILSON, H E, and MANNING, P I (editors). *Memoir of the Geological Survey of Northern Ireland.*

FUNNELL, B M. 1955. An account of the geology of the Bungay District. *Transactions of the Suffolk Naturalists' Society*, Vol. 9, 115–26.

FUNNELL, B M. 1961. The Palaeogene and early Pleistocene of Norfolk. *Transactions of the Norfolk Naturalists' Society*, Vol. 19, 340–364.

FUNNELL, B M. 1972. The history of the North Sea. *Bulletin of the Geological Society of Norfolk*, Vol. 21, 2–10.

FUNNELL, B M. 1893a. The Crag of Bulcamp, Suffolk. *Bulletin of the Geological Society of Norfolk*, Vol. 33, 35–44.

FUNNELL, B M. 1983b. Preliminary note on the foraminifera and stratigraphy of CEGB Sizewell boreholes L and S. *Bulletin of the Geological Society of Norfolk*, Vol. 33, 54–62.

FUNNELL, B M. 1987. Late Pliocene and Early Pleistocene stages of East Anglia and the adjacent North Sea. *Quaternary Newsletter*, Vol. 52, 1–11.

FUNNELL, B M. 1991. Palaeogeographical maps of the southern North Sea Basin; Pliocene (Coralline Crag) to Anglian (Lowestoft Till). *Bulletin of the Geological Society of Norfolk*, Vol. 40, 53–66.

FUNNELL, B M, NORTON, P E, and WEST, R G. 1979. The Crag at Bramerton, near Norwich, Norfolk. *Philosophical Transactions of the Royal Society of London*, Series B, Vol. 287, 489–534.

FUNNELL, B M, and WEST, R G. 1962. The early Pleistocene of Easton Bavents, Suffolk. *Quarterly Journal of the Geological Society of London*, Vol.118, 125–141.

FUNNELL, B M, and WEST, R G. 1977. Preglacial Pleistocene deposits of East Anglia. 246–265 in *British Quaternary studies: recent advances.* SHOTTON, F W (editor). (Oxford: Clarendon Press.)

FYFE, J A, ABBOTTS, I, and CROSBY, A. 1981. The subcrop of the mid-Mesozoic unconformity in the UK area. 236–244 in *Petroleum geology of the continental shelf of North-west Europe.* ILLING, L V, and HOBSON, G D (editors). (London: Heyden and Sons.)

GALLOIS, R W, and MORTER, A A. 1976. Trunch Borehole, Mundesley (132) Sheet. 8–10 in IGS Boreholes 1975. *Report of the Institute of Geological Sciences*, No. 76/10.

GALLOIS, R W, and MORTER, A A. 1982. The stratigraphy of the Gault of East Anglia. *Proceedings of the Geologists' Association*, Vol. 93, 351–368.

GALLOWAY, W E, GARBER, J L, LIU, XIJIN, and SLOAN, B J. 1993. Sequence stratigraphic and depositional framework of the Cenozoic fill, Central and Northern North Sea Basin. 33–43 in *Petroleum geology of Northwest Europe: Proceedings of the 4th Conference.* PARKER, J R (editor). (London: Geological Society.)

GARDNER, K, and WEST, R G. 1975. Fossil ice-wedge polygons at Corton, Suffolk. *Bulletin of the Geological Society of Norfolk*, Vol. 23, 47–53.

GEORGE, M. 1992. *The land use, ecology and conservation of Broadland.* (Chichester: Packard Publishing Ltd.)

GIBBARD, P L. 1988. The history of the great northwest European rivers during the past 3 million years. *Philosophical Transactions of the Royal Society of London*, Series B, Vol. 318, 559–602.

GIBBARD, P L, WEST, R G, ZAGWIN, W H, BALSON, P S, BURGER, A W, FUNNELL, B M, JEFFERY, D H, JONG, J DE, KOLFSCHOTEN, T DAN, LISTER, A M, MEIJER, T, NORTON, P E P, PREECE, R C, ROSE, J, STUART, A J, WHITEMAN, C A, and ZALASIEWICZ, J A. 1991. Early and early Middle Pleistocene subsidence in the Southern North Sea Basin. *Quaternary Science Reviews*, Vol. 10, 23–52.

GODWIN, H. 1978. *Fenland: its ancient past and uncertain future.* (Cambridge: University Press.)

GREEN, C P, and MCGREGOR, D F M. 1990. Pleistocene gravels of the north Norfolk coast. *Proceedings of the Geologists' Association*, Vol. 101, 197–202.

GREEN, C P, MCGREGOR, D F M, and EVANS, A H. 1982. Development of the Thames drainage system in early and middle Pleistocene times. *Geological Magazine*, Vol. 119, 281–290.

GUNN, J. 1867. The order of the succession of the preglacial, glacial and postglacial strata in the coast sections of Norfolk and Suffolk. *Geological Magazine*, Vol. 4, 37–372 and 561.

HALLSWORTH, C R. 1994. Variations in heavy minerals of the Plio-Pleistocene sediments in East Anglia: the implications for provenance. *British Geological Survey Technical Report*, WH/94/88/R

HAMBLIN, R J O, CROSBY, A, BALSON, P S, JONES, S M, CHADWICK, R A, PENN, I E, and ARTHUR, M J. 1992. *United Kingdom offshore regional report: the geology of the English Channel.* (London: HMSO for the British Geological Survey.)

HARLAND, R. 1983. Distribution maps of Recent dinoflagellate cysts in bottom sediments from the North Atlantic and adjacent areas. *Palaeontology*, Vol. 26, 321–387.

HARLAND, R. 1984. Dinoflagellate cyst analyses of samples from the Waveney Valley area, Norfolk. 1-inch Sheet 173. *British Geological Survey Internal Report*, PD2, 84/132.

HARLAND, R. 1993. Dinoflagellate cyst analysis of Quaternary sediments taken from Covehithe and Thorington, Suffolk on 1:50 000 sheets 176 and 191. *British Geological Survey Technical Report*, WH/93/280R.

HARLAND, R, BONNY, A P, HUGHES, M J, and MORIGI, A N. 1991. The Lower Pleistocene stratigraphy of the Ormesby Borehole, Norfolk, England. *Geological Magazine*, Vol. 128, 647–660.

HARMER, F W. 1898. The Pliocene deposits of the east of England. The Lenham Beds and the Coralline Crag. *Quarterly Journal of the Geological Society of London*, Vol. 54, 308–356.

HARMER, F W. 1900a. On a proposed new classification for the Pliocene deposits of the east of England. *Report of the British Association for the Advancement of Science*, Section C, 751–753.

HARMER, F W. 1900b. The Pliocene deposits of the east of England. Part II. The crag of Essex (Waltonian) and its relation to that of Suffolk and Norfolk. With a report on the inorganic constituents of the Crag by Joseph Lomas. *Quarterly Journal of the Geological Society of London*, Vol. 56, 705–744.

HARMER, F W. 1902. A sketch of the later Tertiary history of East Anglia. *Proceedings of the Geologists' Association*, Vol. 17, 416–479.

HARMER, F W. 1904. Field excursion to Cromer, Norwich and Lowestoft. *Proceedings of the Yorkshire Geological Society*, Vol. 15, 305–314.

HARMER, F W. 1910a. The Pleistocene period in the eastern counties of England. 103–123 in *Geology in the field.* MONCKTON, H W, and HERRIES, R S (editors). Geologists' Association Jubilee Volume.

HARMER, F W. 1910b. The Glacial deposits of Norfolk and Suffolk. *Transactions of the Norfolk and Norwich Naturalists' Association*, Vol. 9, 108–133.

HARMER, F W. 1910c. The Pliocene deposits of the eastern counties of England. 86–102 in *Geology in the field.* MONCKTON, H W, and HERRIES, R S (editors). Geologists' Association Jubilee Volume.

HARMER, F W. 1924. The Pliocene molluscs of Great Britain. Vol. 2, Parts I–4. *Monograph of the Palaeontographical Society of London.*

HARMER, F W. 1928. The distribution of erratics and drift. *Proceedings of the Yorkshire Geological Society*, Vol. 21, 79–150.

HART, J K. 1987. The genesis of the North East Norfolk Drift. Unpublished PhD thesis, University of East Anglia.

HART, J K, and BOULTON, G S. 1991. The glacial drifts of northeastern Norfolk. 233–243 in *The Glacial deposits in Great Britain and Ireland*. ELLES, J, GIBBARD, P L, and ROSE, J (editors). (Rotterdam: Balkema.)

HART, J K, HINDMARSH, R C A, and BOULTON, G S. 1990. Styles of subglacial glaciotectonic deformation within the context of the Anglian ice-sheet. *Earth Surface Processes and Landforms*, Vol. 15, 227–241.

HART, J K, and PEGLAR, S M. 1990. Further evidence for the timing of the Middle Pleistocene glaciation in Britain. *Proceedings of the Geologists' Association*, Vol. 101, 187–196.

HAZELDEN, J. 1989. Soils in Norfolk VIII: Sheet TG40 (Halvergate). *Soil Survey Records*, No. 115.

HESTER, S W. 1965. Stratigraphy and palaeogeography of the Woolwich and Reading Beds. *Bulletin of the Geological Survey of Great Britain*, Vol. 23, 117–137.

HEY, R W. 1967. The Westleton Beds reconsidered. *Proceedings of the Geologists' Association*, Vol. 78, 427–445.

HEY, R W. 1976. Provenance of far-travelled pebbles in the Pre-Anglian Pleistocene of East Anglia. *Proceedings of the Geologists' Association*, Vol. 87, 69–82.

HEY, R W. 1980. Equivalents of the Westland Green Gravels in Essex and East Anglia. *Proceedings of the Geologists' Association*, Vol. 91, 279–290.

HEY, R W, and BRENCHLEY, P J. 1977. Volcanic pebbles from Pleistocene gravels in Norfolk and Essex. *Geological Magazine*, Vol. 114, 219–225.

HILL, E. 1902. On the matrix of the Suffolk Chalky Boulder Clay. *Quarterly Journal of the Geological Society of London*, Vol. 58, 179–184.

HOLLINGWORTH, S E, and TAYLOR, J H. 1946. An outline of the geology of the Kettering district. *Proceedings of the Geologists' Association*, Vol. 57, 204–233.

HOLYOAK, D T, IVANOVICH, M, and PREECE, R C. 1983. Additional fossil and isotopic evidence for the age of the interglacial tufas at Hitchin and Icklingham. *Journal of Conchology*, Vol. 31, 260–261.

HOPSON, P M. 1991. Geology of the Beccles and Burgh St Peter district. *British Geological Survey Technical Report*, WA/91/53.

HOPSON, P M, and BRIDGE, D McC. 1987. Middle Pleistocene stratigraphy in the lower Waveney valley, East Anglia. *Proceedings of the Geologists' Association*, Vol. 98, 171–185.

HORTON, A. 1970. The drift sequence and subglacial topography in parts of the Ouse and Nene basin. *Report of the Institute of Geological Sciences*, No. 70/9.

HORTON, A. 1982a. *Geological notes and local details for 1:10 000 sheets TM 17 NW, NE, SW, and SE (Diss, Hoxne, Eye and Occold).* (Keyworth, Nottingham: Institute of Geological Scieneces.)

HORTON, A. 1982b. *Geological notes and local details for 1:10 000 sheets TM 27 NW, NE, and SW (Broderdish, Norfolk and Stradbroke, Suffolk).* (Keyworth, Nottingham: Institute of Geological Sciences.)

INESON, J. 1962. A hydrogeological study of the permeability of the Chalk. *Journal of the Institution of Water Engineers*, Vol. 16, 449–463.

INSTITUTE OF GEOLOGICAL SCIENCES. 1976. Hydrogeological map of northern East Anglia. 2 sheets, 1:125 000. (London: HMSO for Institute of Geological Sciences.)

INSTITUTE OF GEOLOGICAL SCIENCES. 1981. Hydrogeological Map of southern East Anglia. 1:125 000. (London: HMSO for Institute of Geological Sciences.)

JELGERSMA, S. 1979. Sea-level changes in the North Sea basin. 233–248 in The Quaternary history of the North Sea. OELLE, E, SCHUTTENHELM, R T E, and WIGGERS, A J (editors). *Acta Universitatis Upsaliensis Symposia Universitatis Upsaliensis annum Quingentesimum Celebrantis:* Vol.2. (Uppsala.)

JENNINGS, J N. 1952. The origin of the Broads. *Memoir of the Royal Geographical Society*, No. 2, 1–66.

JOLLEY, D W. 1992. Palynofloral association sequence stratigraphy of the Palaeocene Thanet Beds and equivalent sediments in eastern england. *Review of Palaeobotany and Palynology*, Vol. 74, 207–237.

KEMP, R A. 1985. The Valley Farm Soil in southern East Anglia. 179–196 in *Soils and Quaternary landscape evolution*. BOARDMAN, J (Editor). (Chichester: Wiley.)

KEMP, R A. 1987a. Genesis and environmental significance of a buried Middle Pleistocene soil in Eastern England. *Geoderma*, Vol. 41, 49–77.

KEMP, R A. 1987b. The interpretation and environmental significance of a buried soil near Ipswich Airport, Suffolk, England. *Philosophical Transactions of the Royal Society of London*, Series B, Vol. 317, 365–391.

KERNEY, M. 1976. Mollusca from an interglacial tufa in East anglia, with the description of a new species of *Lyrodiscus* Pilsbry (Gastrapoda: Zonitidae). *Journal of Conchology*, Vol. 29, 47–50.

KING, C. 1981. The stratigraphy of the London Clay and associated deposits. *Tertiary Research Special Paper*, No. 6. (Rotterdam: Backhuys.)

KNOX, R W O'B. 1984. Nannoplankton zonation and the Palaeocene/Eocene boudary beds of NW Europe; an indirect correlation by means of volcanic ash layers. *Journal of the Geological Society of London*, Vol 141, 993–999.

KNOX, R W O'B. 1996. Tectonic controls on sequence development in the Paleocene of SE England. 209–230 in Sequence stratigraphy in British geology. HESSELBO, S P, and PARKINSON, D N (editors). Proceedings of the sequence stratigraphy of the Phanerozoic in Britain conference, March 1994. *Geological Society Special Publication*, No. 103.

KNOX, R W O'B, and ELLISON, R A. 1979. A Lower Eocene ash sequence in SE England. *Journal of the Geological Society of London*, Vol. 136, 251–253.

KNOX, R W O'B, and HOLLOWAY, S. 1992. *Lithostratigraphic nomenclature of the UK North Sea*, Vol. 1: Palaeogene of the Central and northern North Sea. KNOX, R W O'B, and CORDEY, W G (editors). (Keyworth: British Geological Survey, for the United Kingdom Offshore Operators Association.)

KNOX, R W O'B, MORIGI, A N, ALI, J R, HAILWOOD, E A, and HALLAM, J R. 1990. Early Palaeogene stratigraphy of a cored borehole at Hales, Norfolk. *Proceedings of the Geologists' Association*, Vol. 101, 145–151.

KNOX, R W O'B, and MORTON, A C. 1983. Stratigraphical distribution of early Palaeogene pyroclastic deposits in the North Sea Basin. *Proceedings of the Yorkshire Geological Society*, Vol. 44, 355–363.

KNOX, R W O'B, and MORTON, A C. 1988. The record of early Tertiary North Atlantic volcanism in sediments of the North Sea Basin. 407–419 in Early Tertiary volcanism and the opening of the NE Atlantic. Morton, A C, and Parson, L M (editors). *Special Publication of the Geological Society of London*, No. 39.

KRINSLEY, D H, and FUNNELL, B M. 1965. Environmental history of quartz sand grains from the Lower and Middle Pleistocene of Norfolk, England. *Quarterly Journal of the Geological Society of London*, Vol. 121, 435–461.

LAMB, H H. 1972. *Climate: present, past and future. Vol. 1: Fundamentals and climate now.* (London: Methuen.)

LAMB, H H. 1977. *Climate: past, present and future. Vol. 2: Climatic history and the future.* (London: Methuen.)

LAMBE, J M, and WHITEMAN, R V. 1979. *Soil mechanics (SI version).* (Chichester: Wiley-Interscience.)

LAMBERT, J M, and JENNINGS, J N. 1960. Stratigraphical and associated evidence. 1–66 *in* The making of the Broads. LAMBERT, J M, JENNINGS, J N, SMITH, C T, GREEN, C, and HUTCHINSON, J N. (editors). *Memoir of the Royal Geographical Society*, No. 3.

LARWOOD, G P, and MARTIN, A J. 1953. Stratigraphy and fauna of the Easton Bavents cliff sections, near Southwold, Suffolk. *Transactions of the Suffolk Naturalists' Society*, Vol. 8, 157–170.

LAWSON, T E. 1982. *Geological notes and local details for 1:10 000 sheets TM 28 NW, NE, SW, and SE (Harleston, Norfolk).* (Keyworth, Nottingham: Institute of Geological Sciences.)

LEE, M K, PHARAOH, T C, and GREEN, C A. 1991. Structural trends in the concealed basement of eastern England from images of regional potential field data. *Annales de la Société Géologique de Belgique*, Vol. 114, 45–62.

LEE, M K, PHARAOH, T C, and SOPER, N J. 1990. Structural trends in central Britain from images of gravity and aeromagnetic fields. *Journal of the Geological Society of London*, Vol. 147, 241–258.

LEE, M K, PHARAOH, T C, WILLIAMSON, J P, GREEN, C A, and DE VOS, W. 1993. Evidence on the deep structure of the Anglo-Brabant Massif from gravity and magnetic data. *Geological Magazine*, Vol. 130, 575–582.

LEES, B J. 1980. Sizewell–Dunwich Banks field study. Topic Report 1 (a) Introduction, (b) Geological background. *Report of the Institute of Oceanographic Sciences*, No. 88.

LEWIS, S G. 1993. The status of the Wolstonian glaciation in the English Midlands and East Anglia. Unpublished PhD thesis, Department of Geography, Royal Holloway, University of London. Abstract: *Quaternary Newsletter*, Vol. 73, 35.

LINSSER, H. 1968. Transformation of magnetometric data into tectonic maps by digital template analysis. *Geophysical Prospecting*, Vol. 16, 179–207.

LITTLE, J A, and ATKINSON, J H. 1988. Some geological and engineering characteristics of lodgement tills from the Vale of St Albans, Hertfordshire. *Quarterly Journal of Engineering Geology*, Vol. 21, 185–199.

LLOYD, J W, HARKER, R, D, and BAXENDALE, R A. 1981. Recharge mechanisms and groundwater flow in the chalk and drift deposits of southern East Anglia. *Quarterly Journal of Engineering Geology*, Vol. 14, 87–96.

LLOYD, J W, and HISCOCK, K M. 1990. Importance of drift deposits in influencing Chalk hydrogeology. 271–278 *in* International Chalk Symposium, 1989. (London: Thomas Telford.)

LONG, P E. 1974. Norwich Crag at Covehithe, Suffolk. *Transactions of the Suffolk Natualists' Society*, Vol. 16, 199–208.

LORD, A R. 1969. A preliminary account of research boreholes at Stradbrooke and Hoxne, Suffolk. *Bulletin of the Geological Society of Norfolk*, Vol. 18, 13–16.

LORD, A R, HORNE, D J, and ROBINSON, J E. 1988. An introductory guide to the Neogene and Quaternary of East Anglia for ostracod workers. *British Micropalaeontological Society Field Guide*, No. 5. 10pp.

LUDWIG, G, MULLER, H, and STREIF, H. 1981. New dates on Holocene sea-level changes in the German Bight. *International Association of Sedimentologists Special Publication*, No. 5, 211–219.

LYELL, C. 1839. On the relative ages of the Tertiary deposits commonly called crag in Norfolk and Suffolk. *Proceedings of the Geological Society of London*, Vol. 3, 126–130.

MALM, O A, CHRIOSTENSEN, O B, ØSTBY, K L, FURNESS, H, LØVLIE, R, and RUSLÅTTEN, H. 1984. The Lower Tertiary Balder Formation : an organogenic and tuffaceous deposit in the North Sea region. 149–170 *in Petroleum geology of the North European margin.* SPENCER, A M (editor). The proceedings of the North European Margin Symposium (NEMS '83) organised by the Norwegian Petroleum Society and held at the Norwegian Institute of Technology (NTH) in Trondheim, 9–11 May 1993. (London: Graham and Trotman Ltd.)

MARTINI, E. 1971. Standard Tertiary and Quaternary calcareous nannoplankton zonation. 39–785 *in* Proceedings of the 2nd plankton conference, Roma 1970. FARINACCI, A (editor). (Rome: Edizioni Tecnoscienza.)

MATHERS, S J. 1988. Geological notes and local details for 1:10 000 sheets TM 45 NW and NE (Iken and Aldeburgh). *British Geological Survey Technical Report*, WA/88/35.

MATHERS, S J, HORTON, A, and BRISTOW, C R. 1993. Geology of the country around Diss. *Memoir of the British Geological Survey*, Sheet 175 (England and Wales).

MATHERS, S J, and ZALASIEWICZ, J A. 1985. Producing a comprehensive geological map. *Modern Geology*, Vol. 9, 207–220.

MATHERS, S J, and ZALASIEWICZ, J A. 1986. A sedimentation pattern in Anglian marginal meltwater channels from Suffolk, England. *Sedimentology*, Vol. 33, 559–573.

MATHERS, S J, and ZALASIEWICZ, J A. 1988. The Red Crag and Norwich Crag formations of southern East Anglia. *Proceedings of the Geologists' Association*, Vol. 99, 261–278.

MATHERS, S J, and ZALASIEWICZ, J A. 1996. A gravel beach-rip channel system: the Westleton Beds (Pleistocene) of Suffolk, England. *Proceedings of the Geologists' Association*, Vol. 107, 57–67.

MATHERS, S J, ZALASIEWICZ, J A, BLOODWORTH, A J, and MORTON, A C. 1987. The Banham Beds: a petrologically distinct suite of Anglian glacigenic deposits from central East Anglia. *Proceedings of the Geologists' Association*, Vol. 98, 229–240.

MERRIMAN, R J. 1983. Composition and origin of glauconite material in Crag deposits from East Anglia. *Proceedings of the Geologists' Association*, Vol. 94, 13–16.

MERRIMAN, R J, PHARAOH, T C, WOODCOCK, N H, and DALY, P 1993. Metamorphic history of the concealed Caledonides of eastern England and their foreland. *Geological Magazine*, Vol. 130, 613–620.

MILLWARD, D, ELLISON, R A, LAKE, R D, and MOORLOCK, B S P. 1987. Geology of the country around Epping. *Memoir of the British Geological Survey*, Sheet 240 (England and Wales).

MITCHELL, G F, PENNY, L F, SHOTTON, F W, and WEST, R G. 1973. A correlation of Quaternary deposits in the British Isles. *Geological Society of London Special Report*, No. 4.

MORTIMORE, R N. 1986. Stratigraphy of the White Chalk of the Sussex coast. *Proceedings of the Geologists' Association*, Vol. 97, 97–139.

MORTON, A C, AND KNOX, R W O'B. 1990. Geochemistry of late Palaeocene and early Eocene tephras from the North Sea Basin. *Journal of the Geological Society of London*, Vol. 147, 425–437.

NEAL, J E. 1996. Summary of Palaeogene sequence stratigraphy in northwest Europe and the North Sea. 15–42 In Correlation of the Early Palaeogene in northwest Europe. CORFIELD, R, and DUNAY, R E (editors). *Geological Society Special Publication*, No. 101.

NICKLESS, E F P. 1971. The sand and gravel deposits of the country south-east of Norwich, Norfolk: Description of 1:25 000 resource sheet TG 20. *Mineral Assessment Report Institute of Geological Sciences*, No. 71/20.

NORTON, P E P, and BECK, R B. 1972. Lower Pleistocene molluscan assemblages and pollen for the Crag at Aldeby (Norfolk) and Easton Bavents, Suffolk. *Bulletin of the Geological Society of Norfolk*, Vol. 22, 11–31.

O'RIORDAN, T. 1980. A case study in the politics of land drainage. *Disasters*, Vol. 4, 393–410.

OWEN, E F. 1970. A revision of the brachiopod subfamily Kingeninae Elliott. *Bulletin of the British Museum of Natural History*, Vol. 19, 27–83.

OWEN, H G. 1971. The stratigraphy of the Gault in the Thames Estuary and its bearing on the Mesozoic tectonic history of the area. *Proceedings of the Geologists' Association*, Vol. 82, 187–207.

PARKER, J M, BOOTH, S K, and FOSTER, S S D. 1987. Penetration of nitrate from agricultural soils into the groundwater of the Norfolk chalk. *Proceedings of the Institution of Civil Engineers*, Vol. 93, 15–32.

PARKER, J M, and JAMES, R C. 1985. Authochthonous bacteria in the Chalk and their influence on groundwater quality in East Anglia. *Journal of Applied Bacteriology*, Symposium Supplement, 15S–25S.

PEAKE, N B, and HANCOCK, J M. 1961. The Upper Cretaceous of Norfolk. *Transactions of the Norfolk and Norwich Naturalists' Society*, Vol. 19, 293–339. Reprinted with addenda in 1970.

PEAKE, N B, and HANCOCK, J M. 1970. The Upper Cretaceous of Norfolk. 293–339 in *The geology of Norfolk*. LARWOOD, G P, and FUNNELL, B M (editors). (London and Ashford: Geological Society of Norfolk.)

PERRIN, R M S, DAVIES, H, and FYSH, M D. 1973. Lithology of the Chalky Boulder Clay. *Nature, London*, Vol. 245, 101–104.

PERRIN, R M S, ROSE, J, and DAVIES, H. 1979. The distribution, variation and origins of pre-Devensian tills in eastern England. *Philosophical Transactions of the Royal Society of London*, Series B, Vol. 287, 535–569.

PHARAOH, T C, MERRIMAN, R J, EVANS, J A, BREWER, T S, WEBB, B C, and SMITH, N J P. 1991. Early Palaeozoic arc-related volcanism in the concealed Caledonides of southern Britain. In *Brussels Caledonide Symposium Volume. Annales de la Société Géologique de Belgique*.

PHARAOH, T C, MERRIMAN, R J, WEBB, B C, and BECKINSALE, R D. 1987. The concealed Caledonides of eastern England: preliminary results of a multidisciplinary study. *Proceedings of the Yorkshire Geological Society*, Vol. 46, 355–369.

PHILLIPS, L. 1976. Pleistocene vegetational history and geology in Norfolk. *Philosophical Transactions of the Royal Society of London*, Series B, Vol. 275, 215–286.

POINTON, W K. 1978. The Pleistocene succession at Corton, Suffolk. *Bulletin of the Geological Society of Norfolk*, Vol. 30, 55–76.

POWELL, A J. 1992. Dinoflagellate cysts in the Tertiary System. 155–251 in *A stratigraphic index of dinoflagellate cysts*. POWELL, J (editor). (London: Chapman and Hall.)

PRESTWICH, J. 1847. On the probable age of the London Clay, and its relation to the Hampshire and Paris Tertiary Systems. *Quarterly Journal of the Geological Society of London*, Vol. 3, 354–377

PRESTWICH, J. 1849. On some fossiliferous beds overlying the Red Crag at Chillesford near Orford, Suffolk. *Quarterly Journal of the Geological Society of London*, Vol. 5, 343–353.

PRESTWICH, J. 1850. On the structure of the strata between the London Clay and the Chalk in the London and Hampshire Tertiary systems. Part I. *Quarterly Journal of the Geological Society of London*, Vol. 6, 252–281.

PRESTWICH, J. 1852. On the structure of the strata between the London Clay and the Chalk in the London and Hampshire Tertiary systems. Part III. *Quarterly Journal of the Geological Society of London*, Vol. 8, 235–264.

PRESTWICH, J. 1854. On the structure of the strata between the London Clay and the Chalk in the London and Hampshire Tertiary systems. Part II. *Quarterly Journal of the Geological Society of London*, Vol. 10, 75–157.

PRESTWICH, J. 1871a. On the structure of the Crag beds of Suffolk and Norfolk with some observations on their organic remains. Part II. The Red Crag of Essex and Suffolk. *Quarterly Journal of the Geological Society of London*, Vol. 27, 325–356.

PRESTWICH, J. 1871b. On the structure of the Crag beds of Suffolk and Norfolk with some observations on their organic remains. Part. III. The Norwich Crag and Westleton Beds. *Quarterly Journal of the Geological Society of London*, Vol. 27, 452–496.

PRESTWICH, J. 1890. On the relation of the Westleton Beds, or Pebbly Sands of Suffolk, to those of Norfolk, and on their extension inland; with some observations on the period of the final elevation and the denudation of the Weald and of the Thames Valley, etc. *Quarterly Journal of the Geological Society of London*, Vol. 46, 84–179.

PRICE, J R. 1980. Acid sulphate and potential acid sulphate soils in the peatlands of the Norfolk Broads. Unpublished report, Soil Survey of England and Wales.

PRICE, M. 1987. Fluid flow in the Chalk of England. 141–156 in Fluid flow in sedimentary basins and aquifers. GOFF, J C, and WILLIAMS, B P J (editors). *Special Publication of the Geological Society of London*, No. 34.

RAWSON, P F, CURRY, D, DILLEY, F C, HANCOCK, J M, KENNEDY, W J, NEALE, J W, WOOD, C J, and WORSSAM, B C. 1978. A correlation of the Cretaceous rocks in the British Isles. *Special Publication of the Geological Society of London*, No. 9.

REID, C. 1882. The geology of the country around Cromer. *Memoir of the Geological Survey*.

REID, C. 1890. The Pliocene deposits of Britain. *Memoir of the Geological Survey*.

RICE, R J. 1968. The Quaternary deposits of central Leicestershire. *Philosophical Transactions of the Royal Society of London*, Series A, Vol. 262, 459–509.

RIDING, J B. 1995a. A palynological investigation of the Easton Bavents Clay and subjacent strata near Southwold (1:50 000 sheet 176). *British Geological Survey Technical Report*, WH/95/9R.

RIDING, J B. 1995b. A palynological investigation of the Kesgrave Beds (Pleistocene) at Thorington, Suffolk (1:50 000 sheet 191). *British Geological Survey Technical Report*, WH/95/41R.

RIDING, J B. 1995c. Additional palynological investigations of the Easton Bavents Clay and subjacent strata near Southwold (1:50 000 sheet 176). *British Geological Survey technical Report*, WH/95/58R.

ROBINSON, A H W. 1980. Erosion and accretion along part of the Suffolk coast of East Anglia, England. *Marine Geology*, Vol. 37, 133–146.

ROSE, J. 1987. Status of the Wolstonian glaciation in the British Quaternary. *Quaternary Newsletter*, Vol. 53, 1–9.

ROSE, J. 1989a. Stadial type sections in the British Quaternary. 15–20 in *Quaternary type sections*. ROSE, J, and SCHLUCTER, C H (editors). (Rotterdam: Balkema.)

ROSE, J. 1989b. Tracing the Bagington–Lillington sands and gravels from the West Midlands to East Anglia. 117–122 in *The Pleistocene of the West Midlands Field Guide*. KEENE, D H (editor). (Cambridge: Quaternary Research Association.)

ROSE, J. 1994. Major river systems of central and southern Britain during the Early and Middle Pleistocene. *Terra Nova*, Vol. 6, 435–443.

ROSE, J, and ALLEN, P. 1977. Middle Pleistocene stratigraphy in south-east Suffolk. *Journal of the Geological Society of London*, Vol. 133, 83–102.

ROSE, J, ALLEN, P, and HEY, R W. 1976. Middle Pleistocene stratigraphy in southern East Anglia. *Nature, London*, Vol. 263, 492–494.

ROSE, J, ALLEN, P, KEMP, R A, WHITEMAN, C A, and OWEN, N. 1985. The early Anglian Barham Soil of eastern England. 197–229 in *Soils and Quaternary landscape evolution*. BOARDMAN, J (editor). (Chichester: John Wiley and Sons.)

SCOTESE, C R, GAHAGAN, L M, and LARSON, R L. 1988. Plate tectonic reconstructions of the Cretaceous and Cenozoic ocean basins. *Tectonophysics*, Vol. 155, 27–48.

SHACKLETON, N J, and OPDYKE, N D. 1973. Oxygen isotope and palaeomagnetic stratigraphy of Equatorial Pacific core V28– 38: Oxygen isotope temperatures and ice volumes on a 10^5 year and 10^6 year scale. *Quaternary Research*, Vol. 3, 39–55, The Middle Pleistocene.

SHENNAN, I. 1989. Holocene crustal movements and sea-level changes in Great Britain. *Journal of Quaternary Science*, Vol. 4, 77–89.

SIESSER, W G, WARD, D J, and LORD, A R. 1987. Calcareous nannoplankton biozonation of the Thanetian stage (Palaeocene) in the type area. *Journal of Micropalaeontology*, Vol. 6, 85–102.

SINCLAIR, J M. 1990. Flint pebbles of northern provenance in East Anglian Quaternary gravels. *Quaternary Newsletter*, Vol. 62, 22–25.

SINCLAIR, J M. 1994. The origin and mineralogy of the Lower Pleistocene Westleton Beds , East Anglia. *Quaternary Newsletter*, Vol. 72, 47–48.

SINHA, A, and STOTT, L D. 1993. Recognition of the Palaeocene/Eocene boundary carbon isotope excursion in the Paris Basin, France (abstract). In *Correlation of the Early Palaeogene in Northwest Europe*, Programme and abstracts from 1–2 December meeting. KNOX, R W O'B, CORFIELD, R, and DUNAY, R E (convenors). Sponsored by the Stratigraphic Committee and The Petroleum Group of the Geological Society of London.

SMITH, A J. 1985. Catastrophic origin for the palaeovalley system of the eastern English Channel. *Marine Geology*, Vol. 64, 65–75.

SMITH, A J. 1989. The English Channel — by geological design or catastrophic accident ? *Proceedings of the Geologists' Association*, Vol. 100, 325–337.

SMITH, C T. 1960. Historical evidence. 63–111 in The making of the Broads. LAMBERT, J M, JENNINGS, J N, SMITH, C T, GREEN, C, and HUTCHINSON, J N (editors). *Memoir of the Royal Geographical Society*, No. 3.

SMITH, N J P (compiler). 1985. Map 1, Pre-Permian geology of the United Kingdom (South). (Keyworth: British Geological Survey.)

SMITH, N J P. 1987. The deep geology of central England: the prospectivity of the Palaeozoic rocks. 217–224 in *Petroleum geology of North-west Europe; proceedings of the 3rd conference, London 26–29 October 1986*. BROOKS, J, and GLENNIE, K (editors). (London: Graham and Trotman.)

SMITH, N J P, JACKSON, D I, ARMSTRONG, E J, MULHOLLAND, T, JONES, S, AULD, H A, BULAT, J, SWALLOW, J L, QUINN, N F, OATES, N K, and BENNETT, J R P. 1985. Map 1, Pre-Permian geology of the United Kingdom (South). Scale 1:1000 000. (Keyworth, Nottingham: British Geological Survey.)

SOIL SURVEY OF ENGLAND AND WALES. 1986. *Annual report*. (Harpenden: Soil Survey of England and Wales.)

SONG, LIN-HAU, and ATKINSON, T C. 1985. Dissolved iron in Chalk groundwaters from Norfolk, England. *Quarterly Journal of Engineering Geology*, Vol. 18, 261–274.

SOPER, N J, WEBB, B C, and WOODCOCK, N H. 1987. Late Caledonian (Acadian) transgression in north-west England: timing, geometry and geotectonic significance. *Proceedings of the Yorkshire Geological Society*, Vol. 46, 175–192.

SPARKS, B W, and WEST, R G. 1968. Interglacial deposits at Wortwell, Norfolk. *Geological Magazine*, Vol. 105, 471–481.

STAMP, L D. 1927. The Thames drainage system and the age of the Strait of Dover. *Geographical Journal*, Vol. 70, 386–390.

STOKER, M S, LONG, D, and FYFE, J A. 1985. The Quaternary succession in the central North Sea. *Newsletters on Stratigraphy*, Vol. 14, 119–128.

STOKER, M S, SKINNER, A C, FYFE, J A, and LONG, D. 1983. Palaeomagnetic evidence for early Pleistocene in the central and northern North Sea. *Nature, London*, Vol. 304, 332–334.

STRAHAN, A. 1913. Boring at the East Anglian Ice Co's works, Lowestoft. 87–88 in *Summary of Progress for 1912*. Geological Survey of Great Britain. (London: His Majesty's Stationery Office.)

STRAW, A. 1960. The limit of the 'Last' Glaciation in north Norfolk. *Proceedings of the Geologists' Association*, Vol. 71, 379–390.

STUBBLEFIELD, C J. 1967. Some results of a recent Geological survey boring in Huntingdonshire. *Proceedings of the Geological Sosiety of London*, No. 1637, 35–44.

SUMBLER, M G. 1983. A new look at the type Wolstonian glacial deposits of Central England. *Proceedings of the Geologists' Association*, Vol. 94, 23–31.

TALLANTIRE, P A. 1953. Studies in the Post-glacial history of the British vegetation. XlII Lopham Little Fen, a Late-glacial site in central East Anglia. *Journal of Ecology*, Vol. 41, 361–373.

TAYLOR, B J, and COOPE, R G. 1985. Arthropods in the Quaternary of East Anglia — their role as indices of palaeoenvironment and palaeoclimate. *Modern Geology*, Vol. 9, 159–185.

TAYLOR, R. 1824. On the alluvial strata and on the Chalk of Norfolk and Suffolk, and on the fossils by which they are accompanied. *Transactions of the Geological Society of London*, Vol. 1, 374–378.

THOMAS, E, and SHACKLETON, N J. 1993. The Palaeocene benthic foraminiferal extinction: timing, duration and association with stable isotope anomalies (abstract). In *Correlation of the Early Palaeogene in Northwest Europe*, Programme and abstracts from 1–2 December meeting. KNOX, R W O'B, CORFIELD, R, and DUNAY, R E (convenors). Sponsored by the Stratigraphic Committee and The Petroleum Group of the Geological Society of London.

TRIMMER, J. 1851. Generalizations respecting the erratic Tertiaries or Northern Drift, etc. *Quarterly Journal of the Geological Society of London*, Vol. 7, 19–31.

TUBB, S R, SOULSBY, A, and LAWRENCE, S R. 1986. Palaeozoic prospects on the northern flanks of the London–Brabant Massif. 55–72 in Habitat of Palaeozoic gas in NW Europe. BROOKS, J, GOFF, J C, and HOORN, B VAN (editors). *Special Publication of the Geological Society of London*, No. 23.

TURNER, C. 1970. The Middle Pleistocene deposits at Marks Tey, Essex. *Philosophical Transactions of the Royal Society*, Series B, Vol. 257, 373–440.

VAIL, P R, MITCHUM, R M, TODD, R G, WIDMIER, J M, THOMPSON III, S, SANGREE, J B, BUBB, J N, and HATFIELD, W G. 1977. Seismic stratigraphy and global changes of sea-level. 42–212 in Seismic stratigraphy, application to hydrocarbon exploration. PAYTON, C E (editor). *Memoir of the American Association of Petroleum Geologists*, No. 26.

VAN STAALDVINEN, C J, VAN ADRICHEM BOOGAERT, H A, BLESS, M J M, DOPPERT, J W, HARSVELT, H M, VAN, MONTFRAM, H M, DELE, E, WERMOUTH, R A, and ZAGWIN, W H. 1979. The geology of the Netherlands. *Meddingen Rijk Geologische Dienst*, Vol. 31, 9–49. (Maastricht: Geological Survey of the Netherlands.)

WARD, D J. 1978. The Lower London Tertiary (Palaeocene) succession of Herne Bay, Kent. *Institute of Geological Sciences Report*, No. 78/10

WEST, R G. 1956. The Quaternary deposits at Hoxne, Suffolk. *Philosophical Transactions of the Royal Society of London*, Series B, Vol. 239, 265–356.

WEST, R G. 1961a. Vegetational history of the Early Pleistocene of the Royal Society borehole at Ludham, Norfolk. *Philosophical Transactions of the Royal Society of London*, Series B, Vol. 155, 437–453.

WEST, R G. 1961b. The glacial and interglacial deposits of Norfolk. *Transactions of the Norfolk and Norwich Naturalists' Society*, Vol. 19, 365–375.

WEST, R G. 1963. Problems of the British Quaternary. *Proceedings of the Geologists' Association*, Vol. 74, 147–186.

WEST, R G. 1980. *The pre-glacial Pleistocene of the Norfolk and Suffolk Coasts.* (Cambridge: Cambridge University Press.)

WEST, R G. 1968. *Pleistocene geology and biology with special reference to the British Isles.* (London: Longmans.)

WEST, R G, and BANHAM, P H. 1968. Short field meeting on the North Norfolk coast. Report by the directors; with an appendix: A preliminary note on the Pleistocene stratigraphy of north-east Norfolk. BANHAM, P H (editor). *Proceedings of the Geologists' Association*, Vol. 79, 493–512.

WEST, R G, and DONNER, J J. 1956. The glaciations of East Anglia and the East Midlands: a differentiation based on stone orientation measurements of the tills. *Quarterly Journal of the Geological Society of London*, Vol. 112, 69–91.

WEST, R G, FUNNELL, B M, and NORTON, P E P. 1980. An Early Pleistocene cold marine episode in the North Sea: pollen and faunal assemblages at Covehithe, Suffolk, England. *Boreas*, Vol. 9, 1–10.

WEST, R G, and GODWIN, H. 1958. The Cromerial interglacial. *Nature, London*, Vol. 181, 1554. WEST, R G, and NORTON, P E P. 1974. The Icenian Crag of south-east Suffolk. *Philosophical Transactions of the Royal Society of London*, Series B, Vol. 269, 1–28.

WEST, R G, and WILSON, D G. 1966. Cromer Forest Bed Series. *Nature, London*, Vol. 209, 497–498.

WHITAKER, W. 1866. On the 'Lower London Tertiaries' of Kent. *Quarterly Journal of the Geological Society of London*, Vol. 22, 404–435.

WHITAKER, W. 1887. The geology of Southwold and the Suffolk coast from Dunwich to Covehithe. *Memoir of the Geological Survey of the United Kingdom*.

WHITAKER, W. 1889. The geology of London and part of the Thames Valley :2. *Memoir of the Geological Survey of England and Wales*. 2 vols.

WHITAKER, W. 1906. The water supply of Suffolk from underground sources. *Memoir of the Geological Survey, England and Wales*.

WHITAKER, W. 1921. The water supply of Norfolk from underground sources. *Memoir of the Geological Survey, England and Wales*.

WHITAKER, W, and DALTON, W H. 1887. The geology of the country around Halesworth and Harleston. *Memoir of the Geological Survey of Great Britain*.

WHITEMAN, C A. 1992. The palaeogeography and correlation of pre-Anglain-Glaciation terraces of the River Thames in Essex and the London Basin. *Proceedings of the Geologists' Association*, Vol. 103, 37–56.

WHITEMAN, C A, and ROSE, J. 1992. Thames river sediments of the British Early and Middle Pleistocene. *Quaternary Science Reviews*, Vol. 11, 363–375.

WILCOX, C J, and HORTON, A. 1982. *Geological notes and local details for 1:10 000 sheets TM38NW, NE and TM39SW, SE (Bungay, Suffolk).* (Keyworth, Nottingham: Institute of Geological Sciences.)

WILLIAMS, W W, and FRYER, D H. 1953. Benacre Ness, an east coast erosion problem. *Journal of the Institute of Chartered Surveyors*, Vol. 32, 772–781.

WILLS, L J. 1951. *A palaeogeographical atlas.* (Glasgow: Blackie and Son.)

WILLS, L J. 1973. A palaogeographical map of the Lower Palaeozoic floor below the Upper Permian and Mesozoic formations. *Memoir of the Geological Society of London*, No. 7.

WILLS, L J. 1978. A palaeogeographical map of the Lower Palaeozoic floor beneath the cover of Upper Devonian, Carboniferous and later formations. *Memoir of the Geological Society of London*, No. 8.

WILSON, B. 1902. *The story of lost England.* (London: George Newnes Limited.)

WOOD, C J. 1988. The stratigraphy of the Chalk of Norwich. *Bulletin of the Geological Society of Norfolk*, Vol. 38, 3–120.

WOOD, S V. 1880. The newer Pliocene period in England. *Quarterly Journal of the Geological Society of London*, Vol. 36, 457–528.

WOOD, S V, and HARMER, F W. 1868. The glacial and post-glacial structure of Norfolk and Suffolk. (Abstract). *Geological Magazine*, Vol. 5, 452–456.

WOOD, S V (JUNIOR), and HARMER, F W. 1877. Observations on the later Tertiary geology of East Anglia. With a note by S V Wood on some new occurrences of mollusca in the Crag and beds superior to it. *Quarterly Journal of the Geological Society of London*, Vol. 33, 74–121. WOODCOCK, N H, and PHARAOH, T C. 1993. Silurian facies beneath East Anglia. *Geological Magazine*, Vol. 130, 681–690.

WOODLAND, A W. 1946. Water supply from underground sources of the Cambridge–Ipswich district. *Wartime Pamphlet of the Geological Survey of Great Britain*, No.20, Part 10.

Woods, H. 1908.   A monograph of the Cretaceous Lamellibranchia of England. Vol. II, part V.   *Palaeontographical Society* [monograph], 181–216, plates 28–34.

Woods, M A. 1993a.   Macrofaunas from the Chalk of the BGS Heath Farm No. 1 Borehole, Halesworth, Suffolk.   *British Geological Survey Technical Report*, WH/93,184.

Woods, M A. 1993b.   The stratigraphy of the Chalk of Sizewell C3 Borehole, Suffolk.   *British Geological Survey Technical Report*, WH/93/312.

Woods, M A. 1994.   Supplementary observations on the Chalk macrofauna of the BGS Heath Farm No. 1 Borehole, Halesworth, Suffolk.   *British Geological Survey Technical Report*, WH/94/83.

Woodward, H B. 1881.   Geology of the country around Norwich.   *Memoir of the Geological Survey (England and Wales)*.

Wyatt, R J. 1981.   *Geological notes and local details for 1:10 000 sheets TM 29 SW and SE (Hempnall and Woodton)*.   (Keyworth, Nottingham: Institute of Geological Sciences.)

Zalasiewicz, J A, and Gibbard, P L. 1988.   The Pliocene to early Middle Pleistocene: an overview.   1–31 in *The Pliocene–Middle Pleistocene of East Anglia*. Gibbard, P L, and Zalasiewicz, J A (editors). (Cambridge: Quaternary Research Association.)

Zalasiewicz, J A, and Mathers, S J. 1984.   Subglacial meltwater channels from the margin of the Anglian ice-sheets in south-east Suffolk.   *Quaternary Newsletter*, No. 44, 46.

Zalasiewicz, J A, and Mathers, S J. 1985.   Lithostratigraphy of the Red and Norwich Crags of the Aldeburgh–Orford area, south-east Suffolk.   *Geological Magazine*, Vol. 122, 287–296.

Zalasiewicz, J A, Mathers, S J, and Cornwell, J D. 1985.   The application of ground conductivity measurements to geological mapping.   *Quarterly Journal of Engineering Geology*, Vol. 18, 139–148.

Zalasiewicz, J A, Gibbard, P L, Peglar, S M, Funnell, B M, Catt, J A, Harland, R, Long, P E, and Austin, T J F. 1991.   Age and relationships of the Chillesford Clay (Early Pleistocene: Suffolk, England).   *Philosophical Transactions of the Royal Society of London*, Series B, Vol. 333, 81–100.

Zalasiewicz, J A, Mathers, S J, Hughes, M J, Gibbard, P L, Peglar, S M, Harland, R, Boulton, G S, Nicholson, R A, Cambridge, P, and Wealthall, G P. 1988.   Stratigraphy and palaeoenvironments of the Red Crag and Norwich Crag formations between Aldeburgh and Sizewell, Suffolk, England.   *Philosophical Transactions of the Royal Society of London*, Series B, Vol. 322, 221–272.

Ziegler, P A. 1981.   Evolution of sedimentary basins in north-west Europe.   3–39 in *Petroleum geology of the continental shelf of north-west Europe*. Illing, L V, and Hobson, G D (editors). (London: Heyden and Sons.)

Ziegler, P A. 1982.   *Geological atlas of Western and Central Europe*.   (Amsterdam: Shell Internationale petroleum Maatschappij BV.)

Ziegler, P A, and Louwerens, C J. 1979.   Tectonics of the North Sea.   7–22 in *The Quaternary history of the North Sea*. Oele, E, Schuttenhelmm, R T E, and Wiggers, A J (editors). *Acta Universitatis Upsaliensis Symposia Universitatis Upsaliensis annum Quingentesimum Celebrantis*, 2. (Uppsala.)

APPENDIX 1

Geological Survey photographs

Photographs illustrating the geology of the Lowestoft–Saxmundham district are deposited for reference in the headquarters of the British Geological survey, Keyworth, Nottingham NG12 5GG; in the library at the BGS, Murchison House, West Main Road, Edinburgh EH9 3LA, and in the BGS Information Office at the Natural History Museum Earth Galleries, Exhibition Road, London SW7 2DE. They belong to the A series and depict details of the various sediments exposed and also include general views and scenery. The photographs can be supplied as black and white or colour prints and 2×2 colour transparancies, at the advertised tariff.

APPENDIX 2

Abstracts of selected borehole logs

Boreholes are listed in alphabetical order and are identified by their numbers in the British Geological Survey registration system. These numbers show the 1:10 000 National Grid sheet followed by the accession number (e.g. TM 39 NE/7).

Halesworth (Heath Farm) Borehole (TM 47 NW/15–16)
[4178 7627]
Drilled by BGS in 1992
Surface level 9.0 m above OD

	Thickness m	Depth m
No recovery	3.08	3.08
CRAG GROUP		
Sand, fine to medium-grained, yellowish brown; clay wisps; shelly below 12.5 m	9.08	13.06
THAMES GROUP (LONDON CLAY)		
Harwich Member		
Mudstone, brownish grey, with several ash layers and cementstones at 13.06–13.10 and 16.85–17.00 m	5.94	19.00
Hales Clay Member		
Mudstone, brown, with pale reddish brown sand laminae; two? ash layers	3.75	22.75
LAMBETH GROUP		
Clay, light bluish grey, strong yellowish brown mottling, passing down at c.23.9 m into thinly bedded yellowish brown sands and grey clays; cementstones at 23.97–24.04 m. Below 25.35 and 25.80 m shelly sand and thin lignitic layers alternate with clay. Below c.25.8 m passes down into lenticular-bedded dark to dusky yellowish brown silty clays and pale yellowish brown sand; at some levels sand from the lenses fills burrows in the underlying clay	7.34	30.09
ORMESBY CLAY FORMATION		
Mudstone, glauconitic, dark greenish grey; locally paler and becoming pale, slightly reddish brown below 35.36 m; bioturbated/burrowed throughout between 34.41 and c.34.65 m. Well rounded flint pebbles at base	6.16	36.25
No recovery	1.87	38.12
UPPER CHALK		
Limestone	2.05	40.17

Hales Borehole (TM 39 NE/7)
[3671 9687]
Drilled by BGS in 1987
Surface level 9.0 m above OD

	Thickness m	Depth m
Uncored	0.65	0.65
HEAD		
Clay, sandy, gravelly	0.43	1.08
CORTON FORMATION		
Till: clay, sandy and sand, part gravelly; poor recovery at 1.44 and 1.65 m	0.63	1.71
KESGRAVE GROUP		
Sand with some layers of gravel including quartz pebbles; poor recovery 2.37 to 2.50, 3.84 to 4.13, 4.76 to 10.00, 10.30 to 12.00 m	10.29	12.00
THAMES GROUP (LONDON CLAY)		
Harwich Member		
Siltstone, sandy, olive-grey; some thin sandstones and many tuffs	4.56	16.56
Hales Clay Member		
Mudstone, silty, pale greyish brown, grading into pale brown sandy siltstone, crudely laminated, and bentonised tuffs	14.27	30.83
Siltstone, pale brown	1.35	32.18
ORMESBY CLAY FORMATION		
Mudstone, dark greyish brown, waxy	1.93	34.11
Mudstone, greyish brown to brownish grey with pale green banding, glauconitic	3.84	37.95
Mudstone, greyish brown to brownish grey, calcareous, glauconitic, alternating with pale grey marl	2.14	40.09
Mudstone, greyish brown, highly glauconitic, calcareous, with fragments of green clay	2.41	42.50
Mudstone, brown with a slight orange tinge down to 45.33 m, pale reddish brown below; poorly bedded, bioturbated	5.00	47.50
Mudstone, brownish grey to greyish brown, silty, noncalcareous, penetrated by burrows (including *Chondrites*) filled with reddish brown clay, from top down to 47.90 m; silty, olive-grey, glauconitic below; phosphatic nodules, calcite concretions, three altered tuffs	10.27	57.77
UPPER CHALK		
Chalkstone with pyritic ghosts of hexactinellid sponges, grading down to white chalk; poor recovery	2.23	60.00

Lowestoft (Lake Lothing) (TM 59 SW/53)
[5380 9260]
Drilled in 1912 (log based on Strahan, 1913)
Surface level 3.7 m above OD

	Thickness m	Depth m
Gravel, sand and clay	21.18	21.18
CRAG GROUP		
Clay, sandy, micaceous	1.37	22.55
Sand and shelly sand	50.60	73.15
THAMES GROUP (LONDON CLAY)		
Clay and sandy clay	48.77	121.92
LAMBETH GROUP (WOOLWICH AND READING BEDS)		
Mottled clay	22.86	144.78
CHALK GROUP		
Chalk and flints	321.56	466.34
?UPPER GREENSAND		
Green clay and chalk, black sand	3.35	469.69
GAULT		
Clay and black sand	13.72	483.41
CARSTONE		
Sandstone, soft and glauconitic	12.50	495.91
?SILURIAN ROCKS		
Shale, slate and sandstone	62.48	558.39

North Warren Borehole (TM 45 NE/11)
[4553 5883]
Surface level 9.7 m above OD

	Thickness m	Depth m
CRAG GROUP		
Sand, greyish orange, fine- to medium-grained, moderately sorted, slightly glauconitic, with silty clay layers up to 10 mm thick	8.0	8.0
CORALLINE CRAG FORMATION		
Aldeburgh Member		
Carbonate sand, dark yellowish orange,		

	Thickness m	Depth m
gravelly, very poorly sorted, coarse-grained. Abundant calcitic shell debris. Aragonitic material absent. Weakly cemented	12.3	20.3
Ramsholt Member		
Carbonate sand, olive and bluish grey, silty, very poorly sorted, medium-grained. Calcitic and aragonitic shells present. Phosphatic pebbles near base	3.3	23.6
LONDON CLAY FORMATION		
Clay, silty, olive-grey	0.7	24.3

South Warren Borehole (TM 45 NE/10))
[4565 5806]:
Surface level 6.4 m above OD

	Thickness m	Depth m
CRAG GROUP		
Sand, dark yellowish orange, medium- to fine-grained, moderately sorted, with rare 10 mm-thick clay layers	5.5	5.5
CORALLINE CRAG FORMATION		
Aldeburgh Member		
Carbonate sand, dark yellowish orange, gravelly, poorly sorted, coarse-grained. Abundant calcitic shell debris. Aragonitic material absent. Patchy weak cement. Phosphatic pebbles near base.	11.75	17.25
Ramsholt Member		
Carbonate sand, greenish grey, poorly sorted, medium-grained. Shelly including aragonitic molluscs. Phosphatic pebbles near base	1.8	19.05
LONDON CLAY FORMATION		
Clay, silty, grey	1.8	20.85

APPENDIX 3

History of survey of the Lowestoft and Saxmundham sheets

The district covered by the Lowestoft (176) and Saxmundham (191) sheets of the 1:50 000 Geological Map of England and Wales was originally surveyed on a scale of one inch to one mile by J H Blake, W H Dalton, C Reid, W Whitaker and H B Woodward, and published on the one-inch scale as Old Series sheets 49NW, 49SW, 50NE, 50SE, 66SE and 67W (Solid and Drift) in 1881–1884.

The district was resurveyed on the scale of 1:10 000 over three periods: first during 1969 by F C Cox as an overlap from the resurvey of the adjoining Norwich (161) sheet; then during 1979–84 by C Wilcox, P M Hopson, A Horton, T E Lawson, A N Morigi, R J Wyatt, S J Mathers and J A Zalaziewicz as part of the Survey's East Anglian Regional Research Project; and finally during 1987–93 by S J Booth, A N Morigi, P S Balson, D H Jeffery, J Pattison, and R J O Hamblin, B S P Moorlock and N G Berridge. The two 1:50 000 scale maps of the district cover both the onshore and the offshore areas; they depict both the Solid and Drift geology.

The following is a list of the 1:10 000 and 1:10 560 scale geological maps included wholly, or in part, in the area of the two 1:50 000 scale maps, Lowestoft (176) and Saxmundham (191), with the initials of the surveying officer(s) and the date of survey for each map.

Manuscript copies of these maps have been deposited for public reference in the library of the British Geological Survey, Keyworth, Nottingham and in the British Geological Survey Information Point at the Geological Museum, Exhibition Road, South Kensington, London. They contain more detail than appears on the 1:50 000 scale maps. Dyeline copies are available for purchase.

Maps, surveyors and dates

TM25NE	Leatheringham	SJB	1992
TM26NE	Dennington	SJB	1991
TM26SE	Framlingham	ANM	1990–1991
TM27NE	Frossingfield	AH	1979
TM27SE	Laxfield	RJOH	1991
TM28NE	Alburgh	TEL	1979
TM28SE	Mendham	TEL	1979
TM29NE	Brooke	FWC	1969
TM29SE	Woodton	RJW	1980
TM35NW	Marlsford	ANM/SJB	1990
TM35NE	Blaxhall	JAZ	1982–1984
TM36NW	Bruisyard	SJB	1991
TM36NE	Yoxford	BSPM	1991
TM36SW	Great Glemham	SJB	1990
TM36SE	Saxmundham	ANM	1991
TM37NW	Linstead Parva	BSPM	1992
TM37NE	Halesworth	SJB	1992
TM37SW	Huntingfield	ANM	1991
TM37SE	Walpole	SJB	1991
TM38NW	Bungay	CW	1980
TM38NE	Ilketshall St Andrew	CW	1980
TM38SW	St James, South Elmham	AH	1980
TM38SE	Ilketshall St Lawrence	RJOH/BSPM	1992
TM39NW	Seething	ANM	1987–1988
TM39NE	Loddon	ANM	1985–1986
TM39SW	Ditchingham	CW	1980
TM39NE	Ellingham	AH	1980
TM45NW	Friston	AJM	1982–1984
TM45NE	Aldeburgh	SJM	1982–1984
TM46NW	Westleton	ANM/BSPM	1991–1992
TM46NE	Minsmere	PSB	1990–1991
TM46SW	Leiston	PSB	1990–1991
TM46SE	Sizewell	PSB	1990–1991
TM47NW	Wenhaston	RJOH	1992
TM47NE	Wangford	BSPM	1992–1993
TM47SW	Thorington	RJOH	1991–1992
TM47SE	Dunwich	ANM	1992
TM48NW	Beccles, south	RJOH	1992–1993
TM48NE	Henstead	SJB	1993
TM48SW	Brampton	NGB/JP	1992
TM48SE	Wrentham	BSPM	1992–1993
TM49SW	Beccles, north	PMH	1984
TM49SE	Burgh St Peter	PMH	1984
TM57NW	Southwold	DHJ/ANM	1992–1993
TM58NW	Kessingland	RJOH	1993
TM58SW	Covehithe	DHJ	1993
TM59SW/SE	Lowestoft	SJB	1993

FOSSIL INDEX

To satisfy the rules and recommendations of the international codes of botanical and zoological nomenclature, authors of cited species are listed below.

Ammonia beccarii (Linné, 1758)   78
Anomia   25
Apectodinium   20
Arctica islandica (Linné, 1767)   25

Baculites sp.   13
Baffinicythere howei (Hazel, 1967)   32, 36
Belemnitella sp.   12
Biflustra savartii (Audouin, 1826)   25
Bison sp.   70
Blumenbachium globosum (Koenig, 1825   25
Bos primigenius Bojanus, 1826   70
Bos sp.   70

Carneithyris   12
Cellaria sp.   25
Cerastoderma edule (Linné, 1758)   40, 78
Chlamys   25
Cliona   25
Coelodonta (*Diceros*) *antiquitatus* (Blumenbach, 1803)   70, 71

Cretirhynchia   12
C. arcuata Pettitt, 1950   13
C. ex gr. *lentiformis* (Woodward, 1833)   12, 13
C. lentiformis?   13
C. woodwardi (Davidson, 1855)   13
Cythere lutea Müller, 1785   36
Cytheropteron nodosalatum Neale and Howe, 1973   32

Desulfovibrio desulfuvicans (Bejerinck, 1895)   80

Elephas antiquus (Falconer & Cautley, 1846)   68
Equus caballus Linné, 1758   70
'*Eschara*' *pertusa* Esper, 1796   25

Finmarchinella logani (Brady and Robertson, 1871)   32

Glycymeris glycymeris (Linné, 1758)   25
Gyropleura inequirostrata (Woodward, 1833)   12

Hydrobia   78
Hyotissa? semiplana (J. de Sowerby, 1825)   12

Kingena   12
Kingena pentangulata (Woodward, 1833)   12

Leptocythere psammophila Guillaume, 1976   32
Lingula   6

Littorina saxatilis (Olivi, 1742)   78
Meandropora aurantium (Milne, Edwards in Lyell, 1838)   25
M. tubipora (Busk, 1859)   25
Mammuthus (*Elephas*) *primigenius* Blumenback, 1803   70, 71
Megaloceros (*Megaceros*) *giganteus* Blumenback, 1803   70

Neithea sexcostata (Woodward, 1833)   12
Nostoceras   13
Nostoceras (*Bostrycoceras*)   13
Operculodinium israelianum (Rossignol) Wall, 1967   36
Orbiculoidea   6
Ostrea   25
Ostrea edulis (Linné, 1758)   78

Phragmites   80, 81, 82
Pontocythere sp.   36
Pseudolimea granulata (Nilsson, 1827)   12, 13
Pseudoptera caerulescens (Nilsson, 1827)   12

Rangifer tarandus (Linné, 1758)   70

Scaphella lamberti (Sowerby, 1816)   25

Thaerocythere mayburya Cronin, 1991   32
Thiobacillus ferroxidans Temple & Colmer, 1951   81
Turbicellepora   25

Zoophycos   13

GENERAL INDEX

See also Contents (p.v) for main headings.

Acadian (early Devonian) orogeny   1
acid soil   81
acid sulphate conditions   80
aeromagnetic anomalies   7
aeromagnetic data   6, 7, 85
Alde valley   78
Aldeburgh   1, 25, 28, 30, 42, 43, 87, 92
Aldeburgh Member   24
Aldeby   66, 87
Aldeby Sands and Gravels   36, 55, 61, 64, 66
Aldringham   47
Alluvial fan deposits   82
Alluvium   82
ammonites   13
Anglian   46, 47, 53
Anglian Region of the National Rivers Authority   88
Anglian Stage   62
Antian   36
aquifers   90
  chalk   89–92
aragonitic fossils, bryozoans   25
aragonitic molluscs   25
artesian conditions   89
Athelington   67
Aurora Formation   37

Bagshot and Claygate Beds   35
Balder Formation   18
bank deposits   79
Barham Arctic Structure Soil   46
Barnby   61
Barnham   53
Barsham Marshes   81
Basal Peat   79
basaltic ash   19
Baventian   36
Beccles   36, 37, 38, 51, 56, 61, 76
'Beccles Beds'   48, 51–52
Beccles Marshes   81
Beeston Chalk   13
Beestonian   46, 47
Benacre Broad   78
Benacre Ness   40, 78, 83
Benacre-Wrentham   61
biozones, local pollen assemblage   68
  Alnus-Corylus   68
  Alnus-Corylus-Quercus   68
  Alnus-Gramineae-Corylus   68
  Alnus-Quercus   68
  Betula-Gramineae   68
  Betula-Pinus-Quercus   68
bivalves   12, 40
Blaxhall   47

blown sand   83
Blyford   72
Blythburgh   78
Blyth estuary   75, 96
Blyth River Gravel Pit   32, 41
Blyth valley   41, 72, 87
boreholes (named)
  Carrow Works   10
  Four Ashes   9
  Hales   15, 16, 21, 86
  Halesworth   15, 18, 19, 20, 21, 23
  Halesworth (Heath Farm)   12, 108–109
  Heath Farm   86
  Lake Lothing   10
  Lowestoft (Lake Lothing)   7, 10, 16, 108
  Ludham   26
  North Warren   26, 109
  Ormesby   16, 19, 20, 31, 37, 86
  Ormesby A   14
  Sizewell   36
  Sizewell C3   12, 15, 18, 21, 23
  South Warren   26, 109
  Stradbroke   28, 29, 30, 31, 36, 40
  Trunch   12
  Wantisden Hall   26
  West Somerton   10
Bottom Bed   *see* Upnor Formation
Bouguer gravity anomaly   7
Bramerton Common   36
Bramertonian   36
Bramertonian transgression   30
Brampton   39
brick clay   87–88
Breydon Formation   53, 75, 76, 79–81, 95
  engineering problems   95
British Eastern Ice Sheet   53, 57
Broadland   1, 74, 75, 76, 79
Broads   88
Broome Beck   66
Broome Heath Pit   70
Broome Terrace   70
Bulcamp   36
Bullhead Bed   15
Bungay   28, 51, 59
Burgh Marshes   71
Burgh St Peter   51, 56, 61
Bytham Sands and Gravels   24, 37, 45, 47–51, 70

calcretes   57
Caledonian orogeny   85
carbonate hardness   93
Carboniferous   4
  structure   85
  tectonism   85
Carstone   10, 109
Catton Sponge Bed   12
Central North Sea Basin   28
Chalk   4, 85, 91
  aquifer   89–92
  Lower   12
  Middle   12
Chalk Group   10
Chalky Boulder Clay   61
Chillesford Brickyard   30

Chillesford Church Pit   32
'Chillesford Beds'   43
Chillesford Clay   26, 30, 33, 42, 87, 93
Chillesford Sand   26, 32, 42
Chippenhall Green   59
chloride concentration   93
chlorite   23
Cimmerian   4, 9
clay mineralogy   23
coastal erosion   95–96
Contorted Drift   54
Cookley   48, 51
Coralline Crag Formation   4, 24–26, 86, 109
Corton   42, 54
Corton Embayment   57
Corton Formation   30, 53–61, 108
  details of localities   59–61
  pebbly sand   57
  sands   54, 56–57
  till   57–58
Corton Woods Sands and Gravels   55, 61, 64, 65
Cove Bottom   39, 87
Cove Bottom Brickworks   88
Covehithe   30, 33, 39, 40, 54, 83, 95
Covehithe Broad   96
Covehithe cliffs   38, 39
Covehithe Ness   83, 96
Crag Group   24, 26–42, 86, 89, 93, 108, 109
  aquifer   93–94
  chronology and biostratigraphy   35–37
  details of localities   35–42
  offshore correlation   37
  stratigraphy and sedimentation   31–35
Crane Formation   37
Crazy Mary's Hole   61
Creeting Beds   26
Creeting Sands   26
Cretaceous   9–13, 85
  Lower   10
  Upper   10–13
  structures   85–86
  tectonic events   85
Cromer Forest-bed Formation   24, 37, 42–44, 50, 61
Cromer Till   54, 55
Cromerian   24, 43, 46

Darsham   41
Devensian   4, 53, 72, 74
dinoflagellate cysts   33
dinoflagellate Zone D5   20
disturbed ground   84
Ditchingham   59
Dogger Bank   74
dune belts   83
Dunwich   40, 41, 95, 96
Dunwich Bank   79
Dunwich Heath   30

Earsham   71, 87
Earth Holes   39
Easton   42
Easton Bavents   30, 33, 36, 40, 96

Easton Broad   96
engineering geology   94–95
eustatic rise of sea level   74

faults
  Bedingfield   28
  Debenham   28
  Pettaugh   28
First Floodplain Terrace deposits   71
Flandrian Stage   4
  transgression   76
flaser bedding   32
flint knappers   87
Flixton   59, 70, 87
Fluvio-marine Crag   26
formations
  Aurora   37
  Balder   18
  Breydon   53, 75, 76, 79–81, 95
  Coralline Crag   24–26, 86, 109
  Corton 30, 53–61, 108
  Crane   37
  Cromer Forest-bed   37, 42, 50, 61
  Harwich   4, 18–21
  Ijmuiden Ground   37
  Ingham   47
  Kidderminster   32, 48
  Lista   18
  London Clay Formation   4, 21–23, 35, 109
  Lowestoft Till   61–67, 87, 94
  Norwich Crag   28, 30, 31, 42, 93
  Ormesby Clay   4, 14–16, 18, 23, 85, 108
  Outer Silver Pit   37
  Sele   16
  Smith's Knoll   37
  Thanet   35
  Upnor   16, 18
  Westkapelle Ground   37
  Yare Valley   53, 71, 72
  Yarmouth Roads   37, 46
Fox Burrows   51
Framlingham   42, 71
'Freshwater Beds'   43

garnets   35
Gault   4, 10, 109
Geldeston   70
geophysical studies   6
Gillingham   61
Gipping till   62
glacial deposits   53–67
glaciofluvial sands and gravels   94
Greensand   10, 109
ground, modified   83–84
groundwater quality   92

Haddiscoe   55
Haddiscoe Sands and Gravels   61, 64, 65–66
Hales Clay Member   18–19, 20, 23, 108
Halesworth   30, 40, 41, 89
Harwich Formation   4, 18–21
'Harwich member'   18, 19–21, 23
Harwich Stone Band   19
head   72
heavy minerals   33

Hedenham   48, 51, 59
Henham Hall   84
Hewett Gas Field   85
high sulphate in groundwater   95
Hills Gravel Member   48
Holocene   74, 77
  Flandrian transgression   74
  valley deposits   79–81
Holton   40, 71, 72
Hoxnian   53
  interglacial   4
  sites in the Waveney valley   68
hydrogeology   88–95

Icenian Crag   26
Ijmuiden Ground Formation   37
Ilketshall St Lawrence   39
illite   23
Ingham Gravel   47
Ingham Sand and Gravel   47
Interglacial deposits   4
  Hoxnian   67–68
  Ipswichian   68
Ipswichian   53
  interglacial period   67
Ipswich–Felixtowe High   14

jarosite   80
Jurassic   4, 85

kaolinite   23
Kesgrave Group   24, 37, 41, 44–47, 108
Kesgrave Sands and Gravels   5
Kessingland   38, 44, 54
Kessingland Level   78
Kessingland–Lowestoft   61
Kidderminster Formation   32, 48
Knettishall   48, 53
Knettishall Gravel   47, 50

Lake Lothing   83
Lambeth Group   4, 16–18, 20, 35, 86, 108, 109
Landscaped ground   83
Latymere Dam   88
Laxfield   40
Leet Hill   51, 59
Leet Hill pit   57
'Leet Hill Sands and Gravels'   55, 59
Leiston   28, 47
Lista Formation   18
lodgement till   62
London Clay Formation   4; 21–23, 35, 109
London–Brabant Massif   1, 6, 7, 10, 85
Lower Chalk   12
Lower Clay   74, 79, 80
Lower Cretaceous   10
  basement   6–8
Lower Peat   74, 81
Lowestoft   36, 38, 54, 65, 67, 83
Lowestoft Denes   78
Lowestoft Ness   78
Lowestoft Till Formation   61–67, 87, 94
Ludhamian   36

made ground   83

magnetic basement   85
magnetic susceptibility   21, 21–23
magnetostratigraphy   16
mammalian remains   70
Mammaliferous Crag   26
marginal meltwater channels   64
Marly Drift   53
Mesozoic   9–13
Metfield   40, 59
Mettingham   48
Middle Chalk   12
Middle Peat   74, 79, 82
Midlands Microcraton   6
mineral deposits   87
Minsmere   78
Minsmere Cliffs   83
Miocene uplift   86
molluscs   78
mudcracks   32

nannoplankton   14
nannoplankton Zone NP9   20
Neogene   24
Newcome Sand   79
nitrate concentration   93
Normanston   65
North Sea Basin, Southern   1, 4, 28, 85
'North Sea Drift'   53
North Warren   43
Norwich Brickearth   54, 57, 58
Norwich Crag Formation   28, 30, 31, 42, 93

offshore correlation   37
Older Blown Sand   73
Orford   36
Orford Spit   78
Ormesby Clay Formation   4, 14–16, 18, 23, 85, 86, 108
Oulton   65
Oulton Beds   55, 64–65
Oulton Broad   83, 88
Outer Silver Pit Formation   37
Outney Common   70, 73, 94
oxygen isotope results   62

Pakefield   38, 44, 54, 95
Palaeogene   4, 13–23, 85
  aquifer   92
  deposits   92
  –Eocene boundary   14
Palaeozoic, Lower   85
  basement   6–8
Pastonian   24, 36, 46
Peasenhall   41
peat   81–82, 88
permeability   89
Permian   4, 85
Pettaugh Fault   86
phosphatised fossils   13
'Plateau Gravels'   65
Pleasure Gardens Till   55, 64
Pliocene transgression   4
pollen assemblage biozones   36
pollen zones   68
porosity   89
postglacial deposits   74–83

Pre-Ludhamian   36
Pre-Pastonian   36, 47
Precambrian   85
proto-Trent   33

quarry   69
  Flixton   38, 59, 69, 70
  Thorington   87
  Wenhaston   87
Quaternary   53, 74
  marine deposits   76–79
Quay Lane Pit   33, 40

Ramsholt Member   25, 109
Red Crag   31, 37, 42, 93
Redisham   61
regional gravity data   7
regional subsidence   85
Reydon   36, 37
Rhaxella chert   33
rip-current channels   33
River Terrace Deposits   68–71
rivers
  Alde   1, 71, 78, 88
  Blyth   1, 68, 71, 88
  Bytham   45
  Deben   1, 71
  Dunwich   41
  Minsmere   88
  Waveney   1, 68, 88
  Yox   1, 43
Rootlet Bed   43, 50

Saint Cross South Elmham   67, 72
Saint Margaret South Elmham   72
saline intrusion   93
sand and gravel   87
'Scandinavian Ice Sheet'   53
Second (Broome) Terrace deposits   70
seismic reflection profiles   6, 78
Sele Formation   16
Seven Hills Gravel   47, 50

shell marl   82
shoreface and beach deposits   76–78
Silurian   89, 109
Sizewell   25, 42
Sizewell Bank   79
Sizewell Member   31
Sizewell–Dunwich Bank   79
smectite   23
Smiths Knoll Formation   37
Snape   42, 47, 64
'soil ripening'   80
Sotterley   61
South Cove   54, 61
South Elmham   39
Southwold   1, 28, 30, 40, 61, 74, 96
Southwold's Town Marshes   78
Starston Till   52, 54, 57, 58, 59
Stradbroke Basin   86
Stradbroke Trough   28, 30, 39
Straits of Dover   74
Sudbury–Bildeston Ridge   7, 86
survey, history   110

Thames Group   4, 18–23, 86, 108, 109
Thanet Formation   35
The Crags   24
The Denes   83
thermal relaxation   85
Third (Homersfield) Terrace deposits   68
Thorington   30, 33, 40, 41
Thorington Pit   33, 35
Thorndon area   86
Thorpe   25
Thorpe Ness   78
Thorpeness   25, 47, 64, 83
Thorpeness Member   31, 36, 37
Thurnian   36
tidal flat deposits   78–79
  environment   32
Timworth   48
Timworth Gravel   47

Toft Monks   47, 48, 50, 57, 61
Toft Monks ridge   43, 51, 57
trace fossils   13
transient electromagnetic (TEM)
  sounding   7, 9, 10, 86
Triassic   4, 85
Type X pollen   68

Uggeshall   39
unconformities   10
Upnor Formation   18
Upper Chalk   12
Upper Clay   79, 80
Upper Cretaceous   10
Upper Greensand   10
Upper Peat   76, 82

Valley Farm Rubified Sol Lessivé   46, 47
Variscan orogeny   85
volcanic activity   14

Waltonian   37
Wangford   40
Wangford Pit   87
water supply   88–95
Waveney valley   1, 30, 37, 38, 44, 47, 48,
  50, 53, 59, 61, 68, 71, 79, 80, 81, 87
Wenhaston   71
Westkapelle Ground Formation   37
Westleton   41
Westleton Beds   26, 28, 30, 32
Westwood Marshes   78
Withersdale   72
Wolstonian   62
Woodbridge   26
Woolwich and Reading Beds   4, 86
worked ground   83
Wortwell   59, 68, 71
Wrentham   39, 54, 87, 94

Yare Valley Formation   71–72
Yarmouth Roads Formation   37, 46

BRITISH GEOLOGICAL SURVEY

Keyworth, Nottingham NG12 5GG
0115 936 3100

Murchison House, West Mains Road, Edinburgh EH9 3LA
0131 667 1000

London Information Office, Natural History Museum
Earth Galleries, Exhibition Road, London SW7 2DE
020 7589 4090

The full range of Survey publications is available through the Sales Desks at Keyworth and at Murchison House, Edinburgh, and in the BGS London Information Office in the Natural History Museum (Earth Galleries). The adjacent bookshop stocks the more popular books for sale over the counter. Most BGS books and reports can be bought from The Stationery Office and through Stationery Office agents and retailers. Maps are listed in the BGS Map Catalogue, and can be bought together with books and reports through BGS-approved stockists and agents as well as direct from BGS.

The British Geological Survey carries out the geological survey of Great Britain and Northern Ireland (the latter as an agency service for the government of Northern Ireland), and of the surrounding continental shelf, as well as its basic research projects. It also undertakes programmes of British technical aid in geology in developing countries as arranged by the Department for International Development and other agencies.

The British Geological Survey is a component body of the Natural Environment Research Council.

Published by The Stationery Office and available from:

The Publications Centre
(mail, telephone and fax orders only)
PO Box 276, London SW8 5DT
Telephone orders/General enquiries 0870 600 5522
Fax orders 0870 600 5533

www.tso-online.co.uk

The Stationery Office Bookshops
123 Kingsway, London WC2B 6PQ
020 7242 6393 Fax 020 7242 6412
68–69 Bull Street, Birmingham B4 6AD
0121 236 9696 Fax 0121 236 9699
33 Wine Street, Bristol BS1 2BQ
0117 926 4306 Fax 0117 929 4515
9–21 Princess Street, Manchester M60 8AS
0161 834 7201 Fax 0161 833 0634
16 Arthur Street, Belfast BT1 4GD
028 9023 8451 Fax 028 9023 5401
The Stationery Office Oriel Bookshop
18–19 High Street, Cardiff CF1 2BZ
029 2039 5548 Fax 029 2038 4347
71 Lothian Road, Edinburgh EH3 9AZ
0870 606 5566 Fax 0870 606 5588

The Stationery Office's Accredited Agents
(see Yellow Pages)

and through good booksellers